IMMUNE

Also available in the Bloomsbury Sigma series:

IMMUNE

HOW YOUR BODY
DEFENDS AND PROTECTS YOU

Catherine Carver

Bloomsbury Sigma
An imprint of Bloomsbury Publishing Plc

50 Bedford Square
London
WC1B 3DP
UK

1385 Broadway
New York
NY 10018
USA

www.bloomsbury.com

BLOOMSBURY and the Diana logo are trademarks of
Bloomsbury Publishing Plc

First published 2017

British Library Cataloguing-in-Publication Data
A catalogue record for this book is available from the British Library.

Library of Congress Cataloguing-in-Publication data has been applied for.

ISBN (hardback) 978-1-4729-1511-5
ISBN (trade paperback) 978-1-4729-1512-2
ISBN (ebook) 978-1-4729-1514-6

2 4 6 8 10 9 7 5 3 1

Chapter illustrations by Matt Dawson
Diagrams by Marc Dando

Bloomsbury Sigma, Book Twenty-seven

Typeset by Deanta Global Publishing Services, Chennai, India
Printed and bound in Great Britain by CPI Group (UK) Ltd, Croydon CR0 4YY

To find out more about our authors and books visit www.bloomsbury.com.
Here you will find extracts, author interviews, details of forthcoming events
and the option to sign up for our newsletters.

Contents

6 CONTENTS

Our Hidden Army

Every epic tale needs a hero. Ours is 12 microns tall and lives for a matter of hours; it's called a neutrophil and your life depends on it. Don't be deceived by its tiny stature and minuscule lifespan; this cell can capture bubonic plague in a web of its own DNA, spew out enzymes to digest anthrax and die in a kamikaze blaze of microbe-massacring glory. The neutrophil is a key soldier in an eternal war between our bodies and the legions of bacteria, viruses, fungi and parasites that surround us. From having sex to cleaning the kitchen sink, everything we do exposes us to millions of potential invaders. Yet we are safe. Most of the time these invaders' attempts are thwarted. This is because the human body is like an exceedingly well-fortified castle, defended by billions of soldiers. Some live for less than a day, others remember battles for years, but all are essential in protecting us. This is the hidden army that we all have inside of us, and I'd like to reveal its myriad of miracles and secrets to you.

Our adventure tour will begin with a jaunt through our key defences, which include tears, snot, stomach acid and a cast of killer cells. Drawing on everything from the history of medicine to cutting-edge science, we'll explore the incredible arsenal that lives within us and how it kills off a plethora of diseases, from the common cold to the plague.

Next we'll rattle through the past, present and future of organ transplants – from the first heart transplant to lab-grown vaginas – and the vital role the immune system plays in their success or failure. While organ transplants are needed by very few of us, the truth is every single one of us is dependent on foreign cells to thrive. This will be our focus in 'Outnumbered' (Chapter 5) as we'll unveil the emerging evidence on the immune system's relationship with the body's billions of resident bacteria.

We'll follow this dirty revelation with some sexy science looking at the role of the immune system in our brief encounters, and maybe even love, as we dance the 'Immunological Tango' (Chapter 6). The chapter that follows, appropriately, will cover pregnancy. Having grown my own cutie of a parasite while writing this book, this chapter is particularly close to my heart as it tells the tale of how the immune system allows, and actually encourages, a baby to thrive, despite being a foreign invader.

Of course not all parasites are cute or precious, and in 'A Palace for Parasites' (Chapter 8) we'll meet the creepy critters that like to call us home and the ways our immune system tries to show them the door. This includes worms that can live in your brain and how one parasite might make you want to skydive. Having come across such unpleasantries it will be a relief to read about our adaptive assassins, our T and B cells, and how we can make as many different antibodies as there are stars in the galaxy. We'll follow this by looking at how we can train these cells to do our bidding as we look at vaccines, a triumph of mankind's manipulation of the immune system that has enabled us to prevent and treat a legion of serious diseases.

However, our story won't all be rainbows and unicorns, as we will also uncover the itchy truth behind allergies and see what happens when the body overreacts to harmless things like strawberries, latex and even exercise. Continuing to cast the immune system as the villain of the piece, we will then investigate its dark side, and see what happens when it turns on the very body it exists to protect. Before we get too down on our defences, we will move to 'Defenceless' (Chapter 13) and dedicate ourselves to understanding the terrible consequences of a weakened immune system and how it makes a normal life near impossible.

From there we will springboard to some of the immune system's most aggressive adversaries. First up is cancer, an opponent worthy of its own chapter as we'll review why the immune system finds it such a difficult challenge and how modern medicine is trying to change that. Then in 'Killer Bugs' (Chapter 15) we will examine Ebola and anthrax – two of the most accomplished killers of the human race – and see why we should be afraid, *very* afraid.

Yet despite all there is to fear, there is also much to be hopeful about. In 'Clever Drugs' (Chapter 16) we will end the book with a future-gazing look at a new dawn of drugs that will harness the power of the immune system and save us from the nightmare of the antibiotic apocalypse.

CHAPTER ONE

Full Frontal Immunity:
The First Line of Defence

Staring at a measuring cylinder full of my own pee, I couldn't help but feel the reality of being a Cambridge student wasn't *quite* how I'd pictured it. I'd definitely imagined less urine and more ancient spires, swarms of bikes and formal dining with Nobel laureates. That class experiment was one of many that showed me the brilliance of our insides (and the desperate lengths a nerd is willing to go to learn). In an era when we spend billions learning about stars trillions of miles away, it's incredible just how little we know about what goes on inside each and every one of us, each and every day. Happily for you, this book offers a collection of curious tales from the immune system without any obligation to harvest your own bodily fluids.

Our story begins with a feat of imagination: if we were to put 100 people in a room, hand them some crayons and ask them to draw a defence system, what might you expect to see? You can have a pretty good guess – probably castles with high, impenetrable walls surrounded by moats (shark-infested, among the more creative participants). Perhaps some soldiers pouring boiling oil on unwanted guests. A less historically inclined artist might draw us an array of lasers, rockets and machine guns. These are relatively predictable because even without knowing what you're defending against, there are certain solid choices you can make. This is akin to the 'innate' arm of the immune system – the set of defences that we are born with, and which essentially remain the same throughout our lives. The innate system is the first line of defence because it's already set up and ready to take on a range of common diseases based on predictable characteristics of the threats to our body. For instance, all invaders need an entry point – it doesn't matter if you're a tiny virus or a massive worm, you need a way in – so part of the innate immune system's role is to have robust control of the body's entry and exit points.

Let's imagine a different task. If we had given our 100 people the challenge of drawing a defence system against a very specific threat, they would have drawn rather different defences. For instance, garlic and holy water would be essential in an anti-Dracula defence system, but would be frankly embarrassing in the face of Darth Vader. This opponent-specific weapons selection resembles the 'adaptive' arm of our immune response, which complements the breadth of the innate response by being able to recognise and respond to the specific threat at hand. Thus our arsenal evolves over our lifetime as we build up a memory of all the infections that we have encountered since the day we were born. So my immune system's bag of tricks might not currently include a smallpox solution, but if I were to contract the disease my adaptive immune response would try its hardest to create one to kill the virus before it killed me. However this adapting process takes time. Thus all my innate defences would essentially hold the fort, and in many instances this first line would be enough to wipe out the invader before the adaptive system gets a chance to craft bespoke weaponry. There are lots of connections between the innate and adaptive immune responses, and the more we discover, the more blurred the boundaries between the two appear. For now though, it's a good way of pulling apart the delightfully complex inner workings of the best defence system on the planet: us.

Skin deep

Our first innate defence is hidden in plain sight: skin. Skin is the largest human organ; if you were to peel yours off you'd lose about 12kg (27lb) instantly. Human skin has been used to make wallets and cover books, and if you were inclined to spread yours out flat it'd make a sizeable rug at about two square metres (22 square feet). Perhaps inspired by this thought, a design company has created a range of furniture that mimics human flesh. Each chubby seat is designed to look, feel and, thanks to pheromone impregnation, *smell* like a hunk of flesh. You can even have a belly button added to it. This may sound strange, but lots of people lounge, snuggle and laze on a genuine sheet of dead-cow skin every evening. Thankfully for the sofa industry they have people much more adept at marketing than I, hence 'leather sofas' not 'dead-cow couches'.

Returning to the plot, the reason we make skin-based sofas is that it's pretty hardwearing stuff. The skin on the soles of your feet is eight times thicker than the skin on your eyelids, but every inch of it is an exquisite barrier that keeps unwanted invaders out. Key to this barrier is the outermost layer, called the epidermis. When we shed skin, it takes the skin cells two to four weeks to migrate through the four layers of epidermis, changing their appearance and function like tiny chameleons as they migrate from the deepest layer up to the surface. The bottom layer is only one cell thick, and it's usually the only point in the epidermis where the cells are capable of dividing – from here on up they're on a one-way journey to death.

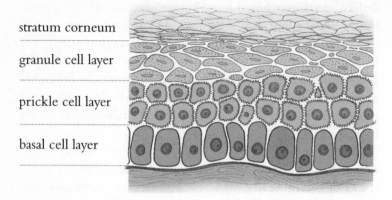

stratum corneum

granule cell layer

prickle cell layer

basal cell layer

Figure 1.1 Cell layers.

The first step on our tiny chameleons' lethal journey involves changing their structure to become the prickle cell layer. It gets its name from the fact that if you look at the cells under a microscope at this stage it seems like they're connected to each other by lots of spines or prickles. The spines are actually keratin fibres and connection points called desmosomes which anchor the cells tightly together to create a relatively impenetrable wall 8–10 cells thick.

As the cells move from this sheet into the granule cell layer above, they begin to lose their nuclei and instead become packed with fat. At the border between this layer and the one above, the cells unleash their fatty payload to waterproof the skin. In a recent quest to understand this waterproofing, scientists from Sweden shaved a sliver of skin from the arms of five volunteers. They then froze the tissue by putting it in a high-pressure freezer at -140°C (-220°F) to capture the exact atomic arrangement in that moment. They then took that Kodak moment and using a chilled diamond blade they sliced the skin into superfine slices and used a microscope cooled to -180°C (-292°F) to look at the structure of the fat layer. What they found was quite unexpected; the arrangement of the fat molecules was unlike anything they'd seen before. Fat molecules usually look a bit like hairpins, with a head and two tails that point in the same direction. However the fat molecules in the frozen skin had tails pointing in opposite directions, as if you took the hairpin and pulled the tails apart to create an almost flat structure. These flattened-hairpin fats stack much more densely than a normal fat layer, creating an even better barrier.

We end at the final resting place of the now dead skin cells: the stratum corneum. The cells here have become flattened stacks of keratin scales, which you slough off and rain down at a rate of roughly 50,000 cells a minute. Now think about the fact that we're all shedding like that, and it's almost unsurprising that dead skin accounts for about a billion tonnes of dust in the atmosphere. Unsurprising, but gross. On the plus side, this constant turnover of cells at the sloughing surface means the barrier is continually replenished and rejuvenated, keeping our skin healthy and keeping the billions of bacteria slathered over its surface out.

Lungs, guts and an accidental shooting

Unfortunately, we can't be truly impenetrable. We need to let in food and water and air and light, and we need to let some things out too. So we have a body full of holes, which is deeply inconvenient from a security perspective. But we have clever holes. Take your mouth: every time you inhale you are sucking about 10,000 bacteria into your lungs. Thankfully your airways are exceedingly well booby-trapped passages lined with goblet cells, which secrete a fine layer of mucus to trap dirt and bacteria. The dirty mucus is then escorted out by microscopic whip-like structures called cilia that stick out from the lining of the airways and beat 1,000–1,500 times per minute, forcing the mucus up and out of the lungs in waves at a rate of 2–3cm/min (~1in/min). Cigarette smoke disables this defence, by slowing, then paralysing and finally killing off the cilia. Smokers are thus left with a lung full of dirty mucus and their body must resort to a less refined method of ejection – the smoker's cough.

While the lung escorts invaders out in an orderly fashion, the gut takes a more medieval approach to border control: acid. The discovery of this acid has a rather gruesome history. The story begins in June 1822 on the Island of Michilimackinac in the wilds of Michigan. At the time the lush green island, christened 'the great turtle' by the Ottawa and Chippewa tribes, was the main trading post of the American Fur Company (the brainchild of America's first multimillionaire, John Jacob Astor). It was while standing in line at the Fur Company store that 20-year-old trapper Alexis St Martin was accidentally shot in the gut at close range. The only doctor on the island arrived to a scene worthy of any horror movie: 'a portion of the Lungs as large as a turkey's egg protruding through the external wound, lacerated and burnt'. St Martin also had fractured ribs, a torn diaphragm and a hole in his stomach through which his breakfast was spilling out on to his shirt. His doctor, an army surgeon by the name of Beaumont, thought St Martin had little chance of survival but endeavoured to help him nonetheless. Astoundingly, St Martin began to recover and with the care of Beaumont he slowly became whole again.

Well, almost. The hole in his stomach didn't fully heal, and St Martin declined offers from Beaumont to stitch it shut, leaving him with a small but permanent connection between the outside world and his stomach. This physical quirk changed not only the course of their relationship but also the history of science. It began on 30 May 1823 when Beaumont's writing first betrayed his experimental interest in St Martin. He described administering medication directly through the hole into St Martin's stomach 'as never medicine was administered to man since the creation of the world'. After this novelty Beaumont took Alexis into his home where he worked as a servant and aide. The good doctor suggested his offer was extended for charitable reasons, and Alexis was likely penniless; however some have since suggested he had less moral (and more ambitious) motives. Centuries have passed, making any judgements on the ethics of the situation speculative. However what we do know is that by 1825 the experiments had begun in earnest, moving their relationship firmly into one of scientist and subject. Beaumont would tie pieces of food to a silk string and introduce them directly through the hole in Alexis' stomach. He tried a range of foods including cabbage, stale bread and salt beef, and also extracted fluid from St Martin's stomach for further analysis.

Over the course of eight years Beaumont conducted 238 experiments on Alexis, eventually publishing the results in a seminal text called *Experiments and Observations on the Gastric Juice and the Physiology of Digestion.* Never before had someone had direct access to the inner workings of a human stomach. According to the founding father of medicine, Dr William Osler, it was Beaumont's research that confirmed that hydrochloric acid was the most important acid in the stomach.

Consequently Beaumont became famous. Alexis St Martin also lived a full life and fathered 20 children before eventually dying at the grand old age of 78. However, while his experiences with the medical profession may have enabled him to live, the actions of his family on his death underline the impact the experience had on him and them. When Alexis died his family refused to have his body buried for four days to ensure it would be too decayed for an autopsy. Aware of the tenacity of the medical profession, and its occasional proclivity for grave robbing, they then buried him under over half a metre (2ft) of rock and almost two metres (6ft) of soil. They expressed the strength of their sentiment through a simple telegram sent to the great Dr Osler himself: 'Don't come for autopsy, will be killed.'

In the years since then, scientists have built on Beaumont's work (with rather more mundane methods) and developed an in-depth understanding of the physiology of stomach or 'gastric' acid. Parietal cells dotted around the surface of the stomach are equipped with proton pumps, which are like tiny merry-go-rounds for ions. They take potassium ions out of the stomach juice and drop them off inside the parietal cell. In a neat little dance of exchange they also pick up hydrogen

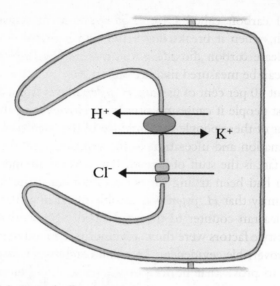

Figure 1.2 Parietal cells moving ions.

ions from inside the parietal cell and release them into the maelstrom of the stomach. These hydrogen ions swirl around in the stomach contents and go on to combine with chloride ions to make hydrochloric acid. This acid is the reason the pH of the normal stomach is an unwelcoming pH 2, capable of disintegrating many of the bacteria that land in it.

One notable exception is *Helicobacter pylori* (aka *H. pylori*), a bacterium that makes its home in the sticky mucus that lines the stomach. While the mucus gives *H. pylori* some protection from the gastric acid, it also employs a bit of clever chemistry to make its home a touch more comfortable. It secretes the enzyme urease, which converts a chemical called urea, naturally present in the stomach, to ammonia. The newly created ammonia is an alkaline, which neutralises the hydrochloric acid in the immediate vicinity of the *H. pylori* and creates a much more inhabitable pH. Historically it was believed that the stomach itself made 'gastric urease' but since the realisation that *H. pylori* was the true chemist, the 'urea breath test' has become a common way of spotting *H. pylori*. Patients swig a drink that contains urea made from a harmless type of carbon called C13. If *H. pylori* is present in the stomach, when it breaks down the C13 containing urea it will release carbon dioxide with an increased level of C13 which can be measured in a breath test.

About 40 per cent of us have *H. pylori* in our stomachs and for most people it causes no problems. However in about 15 per cent of these people the presence of the bacteria leads to inflammation and ulceration in the stomach. The discovery of this fact is the stuff of legend. Barry Marshall and Robin Warren had been trying for years to convince the medical community that *H. pylori* was capable of causing ulcers, but their idea ran counter to the established wisdom that stress and lifestyle factors were the main culprits. In a bid to provide incontrovertible evidence, Marshall underwent a gastric biopsy to prove that he had no *H. pylori* infection and no inflammation of his stomach. He then deliberately swigged some *H. pylori* and had another biopsy two weeks later, which showed that his stomach lining was now inflamed and *H. pylori* ensconced. Combined with all of their previous research, this remarkable feat persuaded the community that Marshall and Warren's radical theory was a genuine breakthrough. This blew the field of ulcer research wide open, leading to over 25,000 research articles, one dedicated journal (*Helicobacter*), a new treatment for stomach ulcers and a Nobel prize for Marshall and Warren.

Tears to fear

Our next defence divides the sexes. The average woman does it 47 times a year, the average man a measly 7 times, and 88 per cent of us feel better afterwards: crying. Yet the tears we shed aren't just cathartic, they're also a form of chemical warfare. Tears contain a mix of electrolytes such as sodium, potassium and chloride, as well as an estimated 80–100 proteins. From this mass of proteins there are three which are found in large quantities (lysozyme, lactoferrin and lipocalin), which help to protect our eyes from the various pathogens we inevitably poke into them with grubby fingers or past-its-best mascara.

So, to the stars of the show. Lysozyme is a minor celebrity in the antibacterial world, known for having been discovered by the legendary Alexander Fleming, king of the accidental scientific find. The year was 1921 and Fleming was on a quest for antibiotics. He had been dropping anything and everything into plates of growing bacteria strewn around his laboratory, hoping to find something to stop them in their tracks. One day he had a cold, and decided to add some of his own mucus to the bacteria. To his great surprise the mucus stopped the bacteria in their tracks, and the reason was lysozyme. This formidable enzyme is found in tears, snot, saliva and cervical mucus. It digests the protective chains of carbohydrate that are key to the make-up of bacterial cell walls. Without its protective wall the bacterium is no longer structurally sound and it bursts at the seams. Alas, lysozyme is too big a molecule to move between cells, preventing it from being an effective way of treating infections, which tend to spread throughout the body. Thus Fleming's quest for a medical antibiotic continued for five years until he came upon his Nobel prize-winning discovery of penicillin.

Lactoferrin has a more humble heritage but it does have the honour of translating roughly as 'milk iron'. What makes it cool is its ability to change the level of iron in blood, milk and other bodily secretions. It was first discovered in milk, hence the name, but we know lactoferrin, like lysozyme, also exists in tears, snot and saliva. It has an incredible propensity to bind iron, and in doing so removes a key mineral that bacteria rely upon to grow. Some bacteria are wise to this and use iron depletion as an indicator that they are inside an animal. Other bacteria have developed their own powerful iron-binding molecules called 'siderophores' which are designed to snatch the iron from the jaws of lactoferrin. Perhaps an even smarter strategy is just to opt out of the iron wars altogether. *Borrelia burgdorferi*, the bacterium responsible for Lyme disease, has replaced iron with manganese as the essential metal element in its enzymes, so an iron–depleted environment is no real threat.

Continuing the immune system tour of the letter 'L', the third key protein in tears is a lipocalin. The proteins of the lipocalin family are an ancient and prolific lot, found not only in humans but in a plethora of other species from rhinos to lobsters. They perform a variety of functions, from pheromone transport to Vitamin A absorption, but the one we're interested in fortifies our tears. About 20 per cent of the protein in our tears is made up of tear lipocalin, whose neat structure includes a pocket for binding a multitude of molecules. This clever pocket allows tear lipocalin to bind the bacterial siderophores we met in the last paragraph, neutralising the bacterium's ability to steal iron from us and leaving it depleted.

Underestimated earwax

Hopefully you've now warmed to the wonder of your own secretions. If not, perhaps I can win you round with another disgustingly cool (and slightly sticky) soldier of the innate immune system: earwax. This humble amber putty is more revealing and intriguing than you'd think. For instance there's evidence it has a distinct smell that varies between races (male Caucasian earwax had higher levels of odorous compounds than that of East Asian men). It's also been used to deconstruct the life story of a blue whale, whose earwax was harvested after death. By analysing the chemical composition along the 24cm (9.4in) plug of earwax, researchers were able to see when in its life the whale had been in polluted water. They were also able to see a spike in the stress hormone cortisol as it reached sexual maturity and faced the challenge of competing for a girl.

Unlike blue whales, we don't retain our earwax. Indeed as a species we are unique in the animal kingdom for many reasons, but our desperate need to get rid of earwax is surely one of the most peculiar. There are hundreds of earwax-removal products on the market; some cost more per millilitre than designer perfume. Others are positively dangerous. Take the 'ear candle', a 25cm (10in) hollow cone made of fabric soaked in paraffin or beeswax, which people put in their ear and set on fire in the mistaken belief it will draw out their earwax (or purify their blood/cure cancer/strengthen their brain. Honestly). In reality ear candles are far more likely to drip hot wax on your face, puncture your eardrum or start a fire. Consequently the US Food and Drug Administration has warned people to steer well clear of them, and has taken to seizing shipments of the dangerous devices.

Putting a burning candle in your ear is a prima facie bad plan, but a more widespread source of harm comes in a far more innocuous shape: the cotton bud. Every year in the UK more people turn up in the emergency department with injuries from cotton buds than from razor blades. Over a third of all British adults continue to jam cotton buds in their ears, despite the fact they are a common cause of hearing loss due to impacted earwax. Perhaps this is because we perceive earwax to be unsightly and dirty, which is why digging it out leaves you feeling clean and oddly satisfied.

However I'd like to suggest this golden goo just has a seriously bad marketing department. Far from being filthy, earwax is an incredible cleaning product that helps solve the ear canal's major logistical challenge: being a cul-de-sac. This anatomical fact means the ear requires a system for escorting any dirt that makes its way into the canal back out the same way. It also means that when the cells that line the canal die, they too must be made to exit – if not they'd pile up, block the canal and leave us deaf. Earwax to the rescue! Earwax is secreted by tiny glands in the skin of the ear canal and includes a mix of cholesterol, triglyceride and another fatty chemical called squalene.* This aural concoction of waxy fatty goodness is further enriched with microbe-killing substances like lactoferrin. Its goopy brilliance traps germs, dirt and dead cells and the whole lot is then carried out of the ear on an earwax conveyor belt. The movement is created by the migration of cells from the eardrum outwards and aided by

---

* Squalene is also found in sharks – it's a key component of shark liver oil, which helps to boost their buoyancy, but is also extracted from the sharks and used for everything from lipstick to haemorrhoid creams.

movements of the jaw during chewing. Thus earwax can catch, kill and kick out the multitude of microbes that wheedle their way into our ears. It might not be pretty, but the next time you reach for a cotton bud or an ear candle remember your ear is busy cleaning itself!

Alas, while snot, skin, stomach acid, sobbing and earwax are all great defences, they're not perfect. Fort Knox allegedly has 30,000 soldiers, an array of armoured helicopters and even laser-triggered machine guns protecting its perimeter. Only a James Bond villain has ever even *tried* to rob it. Sadly the body isn't as secure as Fort Knox, and sometimes bacteria break through the mucus or survive the stomach acid bath, and sneak inside us. But as we'll see in the next chapter, that's where the real immunological fun starts.

CHAPTER TWO

A Cast of Cut-throats: The Killer Cells that Keep Us Safe

The first challenge for our immune system is to spot the intruder. We're made up of an estimated 37 trillion cells, yet just one of these cells is vast compared to a viral particle. A single skin cell is wide enough to fit a queue of 1,000 cold viruses stretching from one side to the other. Bacteria are bigger, but even an average-size bacterium like *E. coli* is only about 1/10th the size of a single human cell. Despite their diminutive size, the immune system is eminently capable of spotting these tiny sneaks, since inside each of us is a surveillance network that would make the NSA green with envy. This network is based on the identification of a series of Pathogen-Associated Molecular Patterns, or PAMPs. PAMPs are structures characteristic of bacteria, viruses and other pathogens that are not found in our cells. Examples include double-stranded RNA (a hallmark of replicating viruses), lipopolysaccharide (part of the outer membrane of bacteria) and unmethylated CpG nucleotides (the building blocks of bacterial DNA). The common theme in these structures is that they are all fundamental to the survival of the pathogens. This is where the beauty of the PAMP lies – because PAMPs are so key to microbial function they are common across many different types of germs and it's exceedingly difficult for bacteria and viruses to change these features to evade detection. Fish gotta swim, viruses gotta replicate and bacteria gotta have cell membranes.

Ninja neutrophils

The immune system is constantly scouring the body for PAMPs using receptors on the surface of a pack of cells which exist to hunt and destroy invaders. This pack is made up of a colourful cast of characters including neutrophils, basophils, macrophages and natural killer cells. First among these is the neutrophil. It defines cool. It's the James Dean of the immune system; it lives fast, dies young and looks good in sunglasses. OK, so the last one has probably never been tested, but I'd put good money on it. The discovery of neutrophils is a quirky tale that can be traced back to the early 1880s. Russian scientist Ilya Mechnikov decided to injure some starfish larvae by stabbing them with tiny thorns from a tangerine tree bought as a Christmas tree for his children. As you do. The next morning he found the thorns surrounded by moving cells, which he hypothesised were consuming and destroying bacteria infecting the injured part of the larvae. This was a momentous discovery that earned Mechnikov a Nobel Prize and introduced the neutrophil to the world.

Initially it was seen as a simple soldier with a basic skill set of 'see bacteria, eat bacteria' but our understanding of the role of the neutrophil has evolved over time. Now we know it is a crafty assassin with a murderous array of killing techniques. Up to 200 million neutrophils gush out of our bone marrow and into the bloodstream every day. They race around the blood on the look-out for evidence of infection. If they find nothing, they die within days. However, in an area of infection neutrophils can be caught by the cells that line the blood vessels, which stick out receptors like tiny hands that loosely grab the moving neutrophils. This slows the neutrophils to a gentle roll along the vessel until they hit another kind of receptor that can stick tightly to the rolling cells, stopping them in their tracks and allowing the neutrophils to migrate through the blood vessel and into the battlefield of the tissue beyond.

This is where the neutrophil's killer skills come to the fore. Unlike regular superheroes, which are boringly confined to one special skill (flying, turning green and shouty, being bat-like, *etc.*) the neutrophil has a range of options at its disposal, from eating its enemy, to catching the microbe in a net spun from the neutrophil's own DNA. It only has to ask one question: which super skill should be deployed for the problem at hand?

This choice can be influenced by other immune system players. For instance, antibodies can bind to the surface of bacteria and in a process called 'opsonisation' make consuming the bacteria more appealing to neutrophils, much like sprinkling tiny chocolate chips on a bacterial cookie. The neutrophil envelops the microbe in an action called phagocytosis or 'cell eating' so that the little invader becomes sealed in a compartment within the neutrophil. The neutrophil can then flood the compartment with enzymes, noxious chemicals and free radicals to destroy the microbe.

In 2004, a previously unappreciated ninja skill of neutrophils was uncovered when Brinkman and colleagues demonstrated that they could create a net spun from their own DNA. Once released, the Neutrophil Extracellular Trap (NET) expands to cover an area up to 15 times the size of the cell that made it, ensnaring any bacteria within its range. The bound bacteria are then swiftly killed and digested by a variety of voracious enzymes studded throughout the nifty NET. Moreover, NETs not only kill bacteria, they also block the spread of infection, stopping the bacteria from making their way further into the body. Centuries ago it was noted that 'good' pus, which resolved infections with haste, usually had a high viscosity. We now know that pus is almost entirely composed of neutrophils and the NETs that give it a thick texture.

Unfortunately, using this weapon seems to come at a cost: the life of the neutrophil. The creation of the NET is thought to be the end result of a programmed self-destruct process called NETosis. Once activated, the changes begin within minutes. First the neutrophil flattens itself. Then over the next hour the shape of its nucleus changes from a multi-lobed splodge to a simple sphere, while the membrane surrounding the nucleus begins to come apart. At the same time the toxic granules within the cell disintegrate to add their poisonous payload to the growing chaos within the neutrophil. Finally, the cell contracts itself tightly before exploding like a party popper that releases deadly NETs instead of streamers.

This account is from the original team who found NETs, but recently there have been other observations of NET creation that contrast with the above, including the possibility that cell death may not be a necessary condition of NET creation. In 2012, a collective of scientists from Canada, France and the US reported that neutrophils do not die immediately after creating NETs. Instead they observed the post-NET neutrophils crawling through infected tissue rapidly and retaining the ability to eat bacteria. The implications of this study are still being debated; scientists are on the steep part of the learning curve when it comes to understanding NETs. The debate is complicated by the fact that much of the knowledge we have is from looking at neutrophils in a petri dish, which is a much simpler environment than the human body. It's a bit like looking at animals in a zoo versus in the wild – you can see a lot of what they get up to but it doesn't mean tap-dancing is normal penguin behaviour. Nevertheless, the fact that NETs can be triggered by all manner of nasties from viruses to fungi and bacteria makes them a potent first line of defence against diseases as diverse as TB and thrush, and a Pandora's box of future learning about the immune system.

A third neutrophil strategy is less elegant but nonetheless effective: spewing microbe-dissolving chemicals into the surrounding tissue. This allows the neutrophil to damage many microbes at once, a bit like fishing by throwing dynamite into the water. However, like dynamite fishing this technique lacks finesse and can cause significant injury to the surrounding tissue. It's thought this neutrophil technique played a part in one of the great mysteries of the 1918 flu pandemic. It was the pandemic that made all those before it look like man flu. One-third of the world's population were infected. It killed more people in one year than the Black Death did in four. It even killed more people than the brutality of the First World War.

Yet the savage death toll of about 50 million souls was not the only remarkable feature of 'Spanish Flu'. It was also unusual because of its victims. Influenza is typically a disease of the elderly and the vulnerable, but this devastating virus was ravaging the young, fit and healthy. Tales from the time describe rapid deaths – women having a girls' night playing bridge into the small hours, only for three of them to be dead the next morning; men walking to work in the morning and being dead by the end of the day. Across an entire population the effect was catastrophic; in the US alone the average lifespan fell by a full decade. At the time there were various theories about the deadly scourge: some believed it was biological warfare emanating from Germany, others apportioned blame more generally to the mustard gas, fumes and smoke of the First World War. Today it is thought to have originated in China, with a mutation in the influenza virus creating a viral coat never seen before. For decades the question of what made this new strain so deadly has been asked by scientists the world over, and the answer is … murky.

The lungs of people who died from Spanish Flu were a pathological conundrum. They were flooded with neutrophils, which begged the question: if such a robust immune response had been launched, then why did they die? Perhaps the neutrophils were part of the problem. Scientists have recently found that in mice infected with the 1918 flu strain an excessive response by neutrophils actually caused destruction of the lung tissue, damaging rather than aiding the sick animals. So it's possible the neutrophils of the young and the healthy in 1918 were too vigorous, and that their friendly fire contributed to the death of the victim. Yet this is not a straightforward tale. Mice with low neutrophil levels have been found to have a higher mortality rate from flu than those with normal levels.

This apparent contradiction is one of many confusing pieces in the 1918 pandemic puzzle. The hunt for an edge piece is driving an ever more adventurous, and some might argue risky, approach to solving the puzzle. In 2005 scientists exhumed the body of a woman who had died in the 1918 pandemic from the Alaskan permafrost. Using samples from her frozen lungs they painstakingly recreated the virus in the hope of finding a definitive answer to the question of why the 1918 pandemic was so deadly. Their experiments have so far pointed to the possibility that the virus jumped from birds to people, and that this meant no person on the planet had any pre-existing immunity to it.

While their quest for the truth continues, some scientists have expressed reservations over the resurrection of the deadliest pandemic in living memory. Perhaps they have in mind the great escapes of another scourge on the human race: smallpox. On 24 August 1978, Mrs Janet Parker, a 40-year-old medical photographer at the University of Birmingham, began to feel unwell. Over the course of the next few days she developed a rash, headache and muscular pains and was eventually admitted to hospital. Her mother also became unwell and both were diagnosed with smallpox. While her mother survived, Janet sadly died.

An enquiry at the time concluded that the most likely source of infection was a smallpox laboratory near to Janet's office. Janet never actually set foot inside the lab but its safety practices were not as rigorous as they ought to have been, allowing a tiny, but deadly, amount of virus to escape into the air. The head of the laboratory was Professor H. Bedwin, a man who had spent a career dedicated to the eradication of smallpox and who was so devastated by the outbreak that he committed suicide before the enquiry finished. This sad tale was not an isolated incident – lab-reared smallpox also caused infections in London in 1973 and in Birmingham in 1966.

One might argue that today's laboratories, guarded with retina scanners and heavily suited lab technicians, are an altogether different kind of prison for a pathogen to escape from. Yet even twenty-first-century labs aren't perfect, as demonstrated by the fact the SARS virus managed to escape on three occasions from high-security labs in east Asia. Nevertheless, given just how deadly 1918 was, perhaps it makes sense to play with fire if it helps stop a future Spanish Flu.

Mega-munching macrophages

Impressive as they are, neutrophils aren't our only cells that eat bacteria. They aren't even the biggest. That honour goes to the macrophage, or 'big eater'. These chunky critters can be found in every tissue in the body: from our brain to our bones, we are riddled with munching macrophages. The majority of macrophages are stationed at strategic sites that provide the best pickings of microbes and dead cells. For instance, the liver is a veritable smorgasbord for the resident macrophages, which dine on aged red blood cells and bacteria freshly delivered from the gut. These liver macrophages are called Kupffer cells in honour of the German scientist who first spotted them, though Kupffer didn't actually realise what they were, confusing the macrophages with part of the lining of the blood vessels of the liver. The realisation that they were distinct cells is attributed to the observant Polish scientist Tadeusz Browicz, who has nothing named after him. Kupffer cells constitute about 80–90 per cent of all tissue-based macrophages in the body, signalling what an important outpost the liver is for the immune system.

Along with the spleen, the liver is the main site for the breakdown of deformed or ageing red blood cells. The Kupffer cells hang around like spiders on the walls of the blood vessels waiting to catch any red blood cells which have passed their best before date (typically 120 days). Once caught, the red blood cell is consumed whole by the Kupffer cell, which sets about dismantling the haemoglobin inside its tasty morsel. Haemoglobin is the protein that allows our red cells to transport oxygen, and it's based on iron, which needs to be released from the red cell back into the blood for recycling.

In the process of freeing the iron from haemoglobin, Kupffer cells also produce a pigment called biliverdin, which in turn is converted to another pigment called bilirubin. These pigments play a key part in the rainbow colours found in bruises. Bruises result from damaged vessels leaking blood into the tissue. This spilled blood rapidly loses the oxygen it was carrying, causing its haemoglobin to change its palette from a vivacious red to a sombre dark blue. Macrophages then arrive to mop up the bodies of dead and dying cells, and the biliverdin released from the haemoglobin breakdown process can be seen in the skin as a greenish tint. As the biliverdin degrades to bilirubin, the bruise takes on the corresponding sallow yellow brown of this final pigment. Eventually the pigment dissipates and the skin looks as good as new.

These bruise-painting macrophages arrive in the tissue thanks to the same siren call of inflammation and infection that beckoned the neutrophils. They tend to arrive later than neutrophils, and on arrival consume not only bacteria but also the spent neutrophil first responders. Sometimes they arrive and do something extraordinary: they become giants. Giant cell formation is a bit of a scientific black box, covered in question marks and emitting an air of mystery. What we do know is that a set of varied and diverse circumstances can prompt multiple macrophages to congregate together and, like a massive Transformer, self-assemble into one magnificent giant cell boasting multiple nuclei. One of the most common types of giant cell is the foreign body giant cell, which forms in response to medical devices, prostheses and foreign objects

too big for a lone macrophage to eat. The giant cell sticks to
the foreign material, and begins to assault the surface directly.
Its size means that the giant cell can cause much more damage
to the implant than a single macrophage, sometimes so much
that the implant fails.

Unleashing giant cells is just one part of a broader immune
response to implants called the foreign body reaction. If the
giant cells fail to annihilate their target, another option for
the foreign body reaction is to wall the implant off from the
rest of the body by surrounding it with a dense fibrous capsule.
This can cause 'implant failure', an all-encompassing term
that can only be defined when you know the primary purpose
of the implant. So, what does failure look like for a breast
implant? Well, they exist to make breasts bigger, so failure is
anything that makes them smaller. Which is exactly what the
fibrous capsule of the foreign body reaction does, squishing,
squeezing and constraining the implant. The fibrous capsule
also frustrates drug-pumps, whose sole purpose is to release a
defined dose of a drug into the body. By encapsulating the
pump, the relatively impenetrable wall stops the pump
delivering its drug payload. These failures have inspired a
wealth of investment in developing new biomaterials. While
there are a few promising candidates, as yet no clinically used
implant is completely immune to the corrosive charms and
fibrous cold shoulder of the foreign body reaction.

Thankfully the gargantuan giant cells have many positive
points to offset their destruction of the perks of breast
implants. One of the most important is the bone-inhabiting
giant cell called the osteoclast, whose role is as the constant
gardener of our bones. Despite their robust nature bones
aren't static structures, they're living leviathans constantly
being broken down and rebuilt by the bony equivalent of yin
and yang: osteoclasts and osteoblasts. The normal-sized
osteoblasts are the builders in this relationship. They arise
from the surface and marrow of the bone and produce the
enzymes and growth factors necessary to construct a sort of
bone scaffolding called bone matrix. This matrix begins to
calcify and harden, trapping the toiling osteoblast in place. Its

task complete, the osteoblast changes into an osteocyte, a bone cell that lives out its days inside the bone and communicates with other bone cells by stretching long finger-like extensions through tiny channels riddled throughout the bone.

While the legacy of the osteoblast is new bone, its counterpart the osteoclast has a more caustic part to play. Osteoclasts spill enzymes and hydrochloric acid on the surface of the bone, doing what giant cells do best: dissolving and destroying. This process of bone resorption removes the old bone to make way for the new, which is vital to bone growth and the repair of fractures. People with genetic disorders that cause their osteoclasts to malfunction develop osteopetrosis, which literally means 'stony bones'. These abnormally dense bones can pinch nerves running through the facial bones and restrict the space for bone marrow to create new blood cells. Moreover, being very dense makes the bones more brittle and therefore prone to fracture. This somewhat counter-intuitive statement is played out when you think about dropping a glass cup versus a plastic one. Glass is more dense than plastic, but which do you expect to break?

Natural born killers

The next cell to take centre stage in our cast of cut-throats is the beautifully named natural killer (NK) cell. Unlike neutrophils and macrophages, which seek and destroy

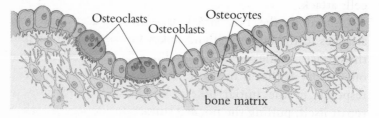

Figure 2.1 Bone with osteocytes, osteoclasts and osteoblasts.

infectious agents roaming our blood and tissues, NK cells target the microbes that have made it *inside* our cells. Though they represent only 1 to 6 per cent of all of our white blood cells, NK cells are distributed far and wide throughout the body because they act as sentinels, interrogating the cells they meet to see if they've been compromised. In practice this usually means cells infected by viruses, because viruses are the squatters of pathogen society. Unlike bacteria, which tend to carry their own internal baggage for all their disease-making needs, viruses pack light. They hold only the genes they need to gain illegal entry to our cells and then instruct our cells' machinery to achieve the virus's aims. The cell provides a very happy home for the virus, and also gives it cover from the immune system.

However, our infected cells betray their inner invaders by sending out distress signals. Some signals are proactive, for example when cells periscope from their surface a receptor called ULBP (UL16-binding protein). Any NK cell that finds itself shaking hands with a ULBP receptor knows it has found a stressed-cell. The same is true if the NK cell extends its receptors to the cell only to find it omits parts of the secret handshake expected from a normal cell. Normal, healthy cells display a range of receptors on their surface which tell the world 'I'm one of us, everything is good'. Touching these receptors placates NK cells, inhibiting their killer ways. Stressed, infected cells display fewer of these normal receptors on their surface and in the absence of their calming presence the trigger-happy NK cells attack.

Central to the NK attack is perforin, a protein that attaches to the target cell and pokes lots of little holes into the surface. While our understanding of perforin is evolving, one theory is that it behaves like a Trojan horse. The cell responds to the initial hole-poking assault by trying to repair itself, pulling the pocked surface inside the cell, then twisting off the damaged part to create a sphere of membrane called a gigantosome that bobs around inside the cell. However, the gigantosome is more than just a pinched-off hole-ridden piece of membrane; its creation was all part of the NK cell's plan to kill the cell. At the same time that it pulls in the perforin holes, the cell unwittingly pulls in a family of protein-eating enzymes called granzymes released by the NK cell. These granzymes don't stay inside the gigantosome for long, seeping out to inflict damage on the cell from the inside. Eventually the gigantosomes themselves break down, spilling any remaining granzymes into the cell, where they trip the cell's self-destruct switch. This process of programmed self-destruction, called apoptosis, brings a neat end to the life of the infected cell. Neat is desirable because it minimises the damage to surrounding cells, which is a risk if a cell dies a messy death retching enzymes everywhere.

Ever neat assassins, NK cells have another method of inducing death that rivals Kill Bill's 'Five Point Palm Exploding Heart Technique' both in name and neatness: TNF (tumour necrosis factor) superfamily-mediated cell death. Best. Name. Ever. And a phrase any mad scientist would be proud to cackle. TNF superfamily-mediated cell

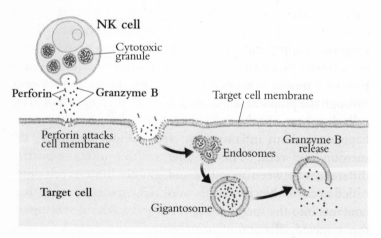

Figure 2.2 Perforin.

death involves a receptor (from the TNF superfamily) on the surface of the infected cell, which is like a magic key capable of unlocking death when turned by the right hand. The death receptor (its actual technical name) is anchored inside the cell by a death domain (yep, also an actual name). When the part of the death receptor exposed on the surface of the cell is stimulated by the NK cell, the death domain unleashes a cascade of chemical events within the cell. One of the most important players in this cascade is a family of enzymes called the caspases. Every cell has many different caspases within it, all of which are kept well under control in an inactive state. However once the first few are activated they play an epic game of tag, recruiting more and more caspases, amplifying each other's actions and driving the cell ever faster towards death. As well as conveying the message to die, the caspases take an active part in dismantling the cell. They rip apart its internal scaffolding, shred the proteins in the nucleus and put a 'cease and desist' on the cell's DNA repair mechanisms. Death is rapid: within hours of the self-destruct command the cell has set its affairs in order, reduced in size and dismembered itself into small packets that can be devoured by macrophages.

The advantage of apoptosis is that it is the most precise form of cell death possible. This precision is beautifully demonstrated during early organism development, when apoptosis sculpts and moulds the growing embryo, carving out a mouse or a kitten or a fish, from a ball of cells. Take the paw of a mouse – it begins like a tiny shovel and it's only through the pinpoint application of death that the intervening cells disappear and the individual digits emerge. This process happens without inflammation and without damaging the surrounding cells. This is one of the most important differences between apoptosis and another kind of cell death called necrosis. Necrotic cells swell and rupture, leaking their contents into the surrounding space and attracting the none-too-targeted affection of neutrophils. Thus chaotic necrotic death is a messy affair, which tends to spread and involve sheets of cells. This is why apoptosis is the preferred execution style for the NK cell, enabling it to target infected cells with minimal collateral damage.

Poking holes in cells, eating bacteria and throwing DNA nets are exceedingly effective strategies, but they're also highly targeted and time-consuming. When it comes to preparing the local area to deal with the invasion we need something far-reaching, we need some chit-chatting chemicals. Let's head to Chapter 3 and hear what they have to say.

CHAPTER THREE

Chit-chatting Chemicals: How Your Immune System Yells

Imagine that a dragon has come to town and you need to get your neighbours to lock their doors and gather their pitchforks. You can either go to each house individually and explain the situation, or run down the street screaming 'DRAGONNNNNNNNNN!!!!' The immune system equivalent of screaming 'dragon' is unleashing chemical messengers that prepare the tissue for dealing with the invaders by setting up a state of acute inflammation and attracting the cadre of killer cells we met in the last chapter.

Naughty nitric acid

The witch's brew of communicative chemicals in our bodies includes nitric oxide (NO), which has quite a reputation, having been named 'Molecule of the Year 1992' by the American Association for the Advancement of Science. NO is produced by macrophages and the lining of blood vessels, and acts as a powerful vasodilator, widening the diameter of blood vessels. By throwing the vessels open, the NO increases the flow of blood to the area and therefore the number of white blood cells arriving on the scene.

This is obviously advantageous during infection, but it's not the only time or place for vasodilation – one of the best is in the bedroom. In 1989 Pfizer began production of a new drug with the catchy name 'U-92, 480', which was capable of increasing the production of NO by blood vessels. They had been seeking a treatment for angina, a disease that causes chest pain due to constriction of the vessels supplying the heart. They began the clinical trial full of hope, but alas the early results were disappointing and they were all set to shut down the trial when volunteers began reporting interesting side-effects: erections. Lots of them. The scientists were understandably keen to investigate further. A model-man was constructed in the lab by taking a test tube filled with inert liquid and dropping a piece of penis from an impotent man inside it. A jolt of electricity sent through the tissue was used to mimic arousal and as expected the blood vessels in the impotent tissue didn't respond. However, when U-92, 480 was added the vessels in the tissue dilated just like in a normal erect penis. And so Viagra was born. It hit the market in 1998 and revolutionised the treatment of erectile dysfunction.

How heat-shock proteins make us stronger

The phrase 'what doesn't kill you makes you stronger' is not just the preserve of Oprah and life-coaches, it's also an accurate description of the biology of heat-shock proteins. When our cells are exposed to heat or cold or infection or a plethora of other stresses, their normal ability to fold proteins into the right shape goes awry. This leaves them unable to craft the proteins they need to work and survive, and produces a build-up of misshapen proteins in the cell. This triggers an action plan called the heat-shock response. One of the key steps in the process is to produce lots of heat-shock proteins (HSPs), which can take misfolded proteins and do the protein equivalent of origami to rearrange the mangled protein into a beautiful, functional shape. Moreover, in doing so the HSPs temporarily transform the cell into a hardened survivor. Experiments have shown that if you expose a cell to a severe, but not quite lethal threat the heat-shock response gears up and allows the cell to withstand subsequent stressors that would normally be deadly.

A symphony of prostaglandins

Arachidonic acid (AA) is a fatty acid found in every cell in the body. Normally it just minds its own business, forming part of the cell membrane. However, when the cell is damaged or threatened, enzymes work to break some arachidonic acid off from the membrane. The acid can then be snapped up by an enzyme called cyclo-oxygenase (COX) that is capable of transforming the fatty acid into a wealth of weird and wonderful molecules including prostaglandins. Prostaglandins take their name from the prostate gland because the first prostaglandin was discovered in semen. Since then we've come to appreciate that prostaglandins are found in a range of places and perform a variety of functions, from preventing blood clots to stimulating contractions during labour.

Prostaglandins (PGs) come in a range of flavours: there's PG E2, PG I2, PG D2 (sadly no R2) and PG F2-alpha. Prostaglandin E2 is a paradoxical prostaglandin – it is pro-inflammatory but also has potent powers to restrain the immune system. It beckons neutrophils, macrophages and other key immune system players to the area, but once they arrive, PG E2 can suppress their killer skills. This may seem counter-intuitive but it's a strategy as sensible as eating bran

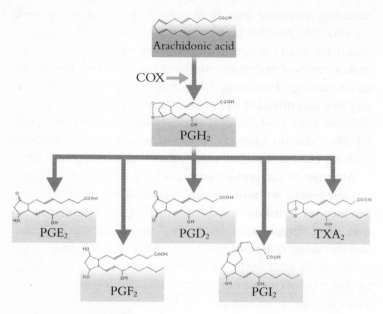

Figure 3.1 Prostaglandins.

flakes for breakfast. Essentially, rather than releasing a pack of really hungry Rottweilers, PG E2 keeps our vicious immune cells on a tight leash, limiting their ability to go on a rampage and unintentionally damage our own cells. It's not the only PG to take this approach – PG I2 and PG D2 also have a mix of pro- and anti-inflammatory effects, which is why PGs have such an important role in orchestrating the immune system symphony.

A storm of cytokines

Our final group of chatty chemicals is also one of the most prolific and challenging to explain. Which is why I felt I needed Batman. Forget Christian Bale's dark, brooding caped crusader, it's all about Adam West's 1960s Batman running around in lycra pyjamas with the words 'POW!' and 'BAM!' appearing during a fight. During one of my favourite episodes Batman arrives on the scene to find a clue left in alphabet soup. Thankfully he packed his alphabet soup collector that morning, allowing him to scoop up the soup and pour it directly into the Batcomputer, which copes with the liquid load better than any laptop I've ever owned. The Batcomputer makes sense of the crazy jumble of letters to decipher the secret message. I mention this because my university lectures on cytokines often felt like wading through an immunological alphabet soup. IL-1, IL-2, IL-6, TNF-alpha, TGF-beta, IFN, GCSF – the list goes on. I spent many hours of revision wishing I had an alphabet soup collector and a Batcomputer...

Attempts to categorise cytokines have been made by many sensible people without a Batcomputer, based on the cells that produce the cytokines or the effects they have, but there are over 40 different examples in the interleukin family alone, and the sheer variety of molecules means every classification system finds itself with a clutch of cytokines that are difficult to shove in any one box. So, rather than rummage through every grouping, let's focus on some of the key players. We'll call this category *The Cytokines' Greatest Hits*: IL-1, interferons and chemokines.

Originally newly discovered cytokines were called interleukins because the first few to be discovered were produced by white blood cells (aka leukocytes), but over time it became apparent that leukocytes weren't the only cell type to make interleukins, which did rather mess up the system. Trying to follow the consequent naming system of the interleukins is like trying to herd some very oddly named cats. One such cat is interleukin-1 (IL-1), which despite the fact that it's often described as a single messenger is actually an entire subfamily of cytokines, consisting of 11 different proteins that help co-ordinate the immediate response to infection. The IL-1 family does this through a range of direct and indirect means, from calling immune cells to the site of infection to provoking cells into secreting other pro-inflammatory messenger chemicals.

The two founding members of the IL-1 family (IL-1 alpha and beta) are mainly made by macrophages and are included in my *Cytokines' Greatest Hits* for their ability to make us hot. These pro-inflammatory little fire-starters travel in the blood to the base of the brain where they reset our internal thermostat, causing the body to reach a feverish temperature. This creates a hostile environment for bacterial and viral infections, which are temperature-sensitive and designed to thrive in the normal human body climate of about 37°C (98.6°F). A higher temperature is relatively manageable for our cells, but if it climbs above about 40.5°C (105°F) our own proteins can start to lose their shape and function. This can cause seizures, collapse and even death, which is why doctors sometimes resort to rapid cooling techniques for patients with extremely high body temperatures.

The fastest of the rapid cooling techniques is evaporative cooling, which involves removing the person's clothing and misting them using spray bottles filled with tepid water while large fans point at the person and circulate the room air. This leads to super-rapid cooling, reducing the person's core temperature by about 0.3°C per minute, which is why the plan has to be accompanied by continuous monitoring using a rectal thermometer. This may sound unpleasant, but perhaps less so when compared with the 'whole-body ice-packing' experience. This necessitates lying naked on a plastic sheet or in a child's paddling pool and being covered in crushed ice. On the plus side the ice remains external, unlike invasive gastric cooling, which is best left to the imagination.

Our next clever chemical messengers are the interferons, which are a group of proteins that do what they say on the tin – interfere. Interferons are made by many different cells including macrophages and NK cells, and are triggered by the presence of viral RNA. The interferon molecule insinuates itself into the local area and binds to receptors on the surface of cells, allowing it to switch on hundreds of genes inside the tissues it touches. These genes cause the cells to make proteins that inhibit viral replication in infected cells and fortify uninfected cells so they can repel any viral particles that try to gain entry. Interferons also activate any nearby NK cells, making them alert to the presence of a virus in their neighbourhood. Interferons are so effective at interfering that pharmaceutical companies have made synthetic versions, which are used to treat hepatitis C. Prior to the advent of synthetic interferons we had no effective treatment for this serious liver disease, making synthetic interferons a rare success story in our attempts to make drugs that can treat viral infections.

The third noteworthy set of cytokines in our whistle-stop tour of this massive family of messengers are the chemokines. More than 50 different chemokines have been discovered to date, and the list just keeps getting longer, making them the biggest subtype of cytokine known to science. The chemokines are the Pied Pipers of the cytokine world, united as a group by their ability to beckon a colourful array of cells to a particular location. In a process known as chemotaxis, they can call up neutrophils, macrophages and other immune system soldiers to mount a response to injury and infection and begin the healing process.

The cytokines we've just discussed, along with the multitude we haven't mentioned, form an exceedingly efficient internal mail system that communicates immune system messages throughout the body. However, some diseases can cause this system to go into overdrive, serving the body with an overwhelming cytokine spamming. This cytokine-induced chaos is known as a 'cytokine storm' and can be caused by an array of infectious and non-infectious agents, from blood transfusions to SARS. The cytokine storm was first described in 1993, but really rose to prominence in 2005 when its role in making bird flu more vicious than ordinary flu meant the term was being discussed by the scientific community and the public alike. Now it has had a progressive metal band and an amateur Irish football team named after it, and has even acquired its own hashtag on Twitter. Yet like many things celebrated in the social-media sphere, the cytokine storm is notorious for less than wholesome reasons: it kills people.

This reality was brought into stark relief in 2006 when a clinical trial of a new cancer drug went terribly wrong. The drug, TGN1412, was supposed to jump-start the immune system of leukaemia patients whose disease had decimated a type of white blood cells called T cells. Normally T cells need two separate on-switches to be flicked at the same time to activate their invader-destroying skills. Animal studies had shown TGN1412 could subvert this system and activate T cells without the need for any other signals. It was hoped this ability to maximise T cell activation would compensate for the smaller numbers of T cells in the leukaemia patients and therefore improve their ability to fight infections. The drug was in the very early stages of testing in people, known as a Phase I trial, the purpose of which is to test the safety of a drug in healthy people. Six healthy young men were given TGN1412 and two were given an inactive placebo. Within 60 minutes of TGN1412 being pumped into their bodies the men developed headaches, muscle pains, diarrhoea, rising heart rates and falling blood pressures. They were showing all the signs of a cytokine storm. Over the next few hours their number of neutrophils soared, their cytokine levels went higher than any test could measure and their organs began to fail. The massive over-activation of their immune system by TGN1412 was killing them. The six men were rushed to intensive care where they were pumped full of steroids to suppress their potentially lethal immune response. Their failing kidneys were supported by dialysis and their injured lungs were mechanically ventilated. None died but all were brought to the brink of death by a cytokine storm.

If you're wondering why all the people in the clinical trial were male, you should be. Historically men have been the clinical trial participant of choice, but in recent times there has been a push to redress the balance as it has become apparent that there are differences in how the sexes experience illness and respond to drugs. In 1993 the National Institutes of Health in the US required the inclusion of women and minorities in any research funded with their dollars, which has helped, but there is still a long way to go. Take cardiovascular disease, one of the biggest killers in the Western world; its impact on men and women is different in almost every way, from the physiology of the disease to the symptoms it causes. However only a third of people in cardiovascular disease clinical trials are female and less than 31 per cent of the trials separate out their results by sex. This makes it exceedingly difficult to detect differences in the way the sexes respond to an intervention, leaving both sexes worse off. Women are particularly poorly served by this inequity in numbers because many of the drugs prescribed by doctors or sold over the counter have never been tested in large numbers of women prior to release.

A deadly complement

So, as we've seen, our blood is much more than a simple transport medium for red and white blood cells; it's actually teeming with chemicals that can tell these cells where to go and what to do. These crafty chemicals can escalate inflammation and drive us to fever point, or quench our excitable immune system by telling our soldier cells to kick back and take some R&R. The two systems – cells and chemicals – complement each other. This is never truer than in the case of our next chemical character, which is why it has the well-deserved moniker 'the complement system'.

The complement system comprises over 30 proteins found in the blood and on the surface of our cells. When it comes to the names of these proteins and their chemical pathways, you'd be forgiven for wondering whether one of the lab monkeys chaired the meeting. For instance, the classical pathway, the first of three complement pathways to be discovered, involves sequential activation of complement proteins in the following order: C1, C4, C2, C3, C5. The random number generator stops after this with a simple C6, C7, C8 and C9 following. This naming pattern is a relic of the order of scientific discovery, which is also why the second pathway to be discovered was called the alternative pathway. It must have been an awkward moment when someone realised there was a third pathway. The 'alternative alternative pathway' was never going to catch on, and so the third was called the lectin pathway after the carbohydrate-binding protein that activates it.

In fact each pathway is activated in a different way, ensuring lots of different triggers can bring about the triad of goals of the complement system: defence against the dark arts of pus-producing bacteria, acting as a bridge between the innate and adaptive arms of the immune response, and cleaning up the detritus of inflammation once the healing begins.

The classical pathway is tripped when C1 is activated by coming into contact with bacteria or by the smoke signals of infection, such as DNA escaping from dying cells. This causes C1 to break apart, unleashing its active form, which goes on to start a cascade of chemical reactions fizzing away. The alternative pathway works slightly differently. It's based on a constant low-level spontaneous activation of C3 in the blood, breaking it up into C3a and C3b and scattering the C3b like a showering of tiny hand grenades on the surrounding cells. Our own cells have the ability to disarm these grenades but bacteria do not. The lectin pathway is another one that is triggered by the presence of bacteria, but can also be tripped by our own cells if they are dying or cancerous.

While the three pathways are activated in different ways, they all converge in a final common path that produces a weapon with a name worthy of a Bond villain's invention: the Membrane Attack Complex. The Membrane Attack Complex (MAC) is created when several complement proteins come together and form a pore that is inserted into the target cell. This leaves the target (usually bacteria) incapable of controlling the entry or exit of water, which is deadly. This technique is particularly effective against one of the bacteria responsible for meningitis: *Neisseria meningitidis*. Once *N. meningitidis* has made its way inside us, it sets up shop inside our cells, which is unusual because bacteria normally live outside cells. One of the most important techniques for overcoming this particular invader is to kill it before it breaks in, which is exactly what the MAC does. The importance of this fact was demonstrated by a study of Japanese people with genetic disorders that damaged their ability to create the Membrane Attack Complex. Those affected were at a 5,000 times higher risk of developing meningococcal disease (a bacterial form of meningitis) than Japanese people with normal complement systems.

Genetic studies have also demonstrated the importance of another complement skill: sprinkling C3b on the surface of bacteria makes them much more appetising to microbe-munching cells like macrophages and neutrophils. People with genetic mutations that reduce complement's ability to fulfil this seasoning role suffer from repeated pus-heavy infections in childhood. As a consequence many are afflicted with a 'failure to thrive', which means their rate of growth tails off compared with other children of the same age. Sounds pretty negative, right? It's therefore a bit of a surprise to discover that one of these genetic mutations is common among people from certain ethnicities. This suggests we're missing part of the puzzle and that there is some benefit to this mutation that outweighs the cost incurred in childhood. One theory is that it protects people against infection by mycobacteria, the family of bacteria that causes TB and leprosy. Mycobacteria are a bit of an oddity: they actually *want* to be gobbled up by our macrophages because that's where they set up home and cause disease. Thus it's possible that complement's ability to make these bacteria seem tastier to our macrophages plays right into the mycobacteria's pili.[*] Supporting evidence for this theory comes from a study from Ethiopia, which found that people infected with leprosy were more likely to have high levels of a key complement system protein than their healthy counterparts. Moreover, other studies have shown that mycobacteria are capable of making their own C4 copycat molecule that triggers our complement cascade.

All of the elements of the immune system we've met so far, from the complement cascade to ninja neutrophils, have been

---

[*] The closest thing to hands a bacterium has.

honed over thousands of years to attack foreign invaders. One thing evolution didn't bargain on was us developing the ability to extract organs and transplant them into other people. The success or failure of a transplant is dependent on the reaction of the immune system, and it is this world of complex immunology, messy medical ethics and designer vaginas that we head off to explore next.

CHAPTER FOUR

Transplants: Designer Vaginas to Pig Heart Perfection

Paddling across a river in Tokyo is a non-standard use of a vagina. That didn't stop Japanese artist and agent provocateur Megumi Igarashi, who made the journey across the river in 2013 in a kayak based on a 3D model of her own genitalia. This aquatic endeavour resulted in Igarashi being arrested for the distribution of 'obscene' data, because she allegedly offered to share the computer code for the design with the biggest backers in her crowdfunding campaign, allowing them to replicate the kayak. The whole process was made possible thanks to 3D printing, as a jubilant Igarashi tweeted to Lady Gaga: 'My vagina was scanned by the 3D printer and it was expanded to 2-meter size as the kayak.'

That is one way to make a vagina, but as discussed, for a fairly non-standard use. To make a real, all-singing-and-dancing vagina takes more than a 3D printer, but it has been done. In 2014, doctors from Mexico and the US published a paper in the *Lancet*, reporting the successful transplantation of lab-grown vaginas into four young women affected by Mayer-Rokitansky-Küster-Hauser Syndrome (MRKHS). This syndrome affects up to 1 in 1,500 female births, and causes the girls to be born with a wholly or partly absent vagina. Girls are usually diagnosed in their early teens when their periods fail to appear, and the standard treatments are dilation, progressive traction and surgery. Ouch.

Over the years many different types of synthetic vagina have been carefully crafted for girls with MRKHS, using everything from cellulose (a major component of plant cell walls) to minced-up buccal mucosa (the lining of the mouth). The team in our story did something a bit different: they took cells from each girl's vulva (external genitalia) and got these cells to grow on a biodegradable scaffold that was designed to be the perfect fit for that girl. Like a sci-fi Cinderella's glass slipper in vagina form. After an average of 6.75 years of follow-up, all four girls gave their designer vaginas ratings well within the normal range for lubrication, arousal, orgasm, satisfaction and absence of pain. In short, the transplants were a success.

Part of what made these transplants a success was the fact that they were made from the girls' own cells. Therefore the immune system saw the cells of the brand spanking new vagina as familiar old friends in a slightly different venue, and didn't reject them. This immunological acceptance also holds for other self-to-self or 'autologous' transplants, a fact that has been exploited in a range of ways. For instance, having a thumb is *much* more useful than having a big toe. Many of the fine, complex movements we do without thinking, like holding a pen to write or snapping our fingers, would be incredibly difficult without a thumb – you're effectively left with a hand whose only action is to clamp. In contrast, having a big toe go AWOL may make balance a little more difficult (and certainly makes wearing peep-toe shoes an interesting conversation starter) but that's about it. Therefore some people who lose their thumb elect to have one of their big toes transplanted onto their hand as a replacement, creating what is sometimes known as a 'thoe'. The thoe requires a bit of getting used to, and the procedure itself takes an immense amount of skill (one surgeon compared it to 'sewing together a chopped chive with a human hair') but immunologically speaking it's a snap.

The penis, while a simple creature in many ways, is a much harder organ to transplant. Unlike the vagina, we can't grow a penis in a lab, and clearly a big toe isn't going to perform as a replacement. So, it's to cadavers that the perseverant penis transplant surgeon must turn. In 2006 the world's first penis transplant was conducted on a Chinese man who had been in an accident that left him with a 1cm stump for a penis. Unable to urinate normally or have sex, he volunteered to be the first man to undergo this complex and delicate operation. And so it came to pass that after 15 hours in theatre the 44-year-old man woke up with a 20-something penis. Much like the toe/thumb, the operation to attach the 10cm (4in) appendage required an immense amount of skill in meshing the two sets of blood vessels, skin and sinew together. Little did these master craftsmen know they'd be lopping it off again in two weeks' time.

Initially the operation was hailed a success, with the penis gaining a good blood supply and enabling the man to urinate normally through it. However the sense of success was short-lived and it was ultimately rejected – not by the immune system, but by the man and his wife. In some ways it sounds bizarre – bearing in mind the lengths they had gone to, the risks they had taken on – to then change their minds after two weeks and request its removal. On the other hand, it's hard to overestimate the psychological 'ick' factor. It doesn't require *such* a stretch of the imagination to appreciate that the reality of living with a dead man's penis would be... challenging.

Why your mind matters

This 'ick' phenomenon isn't unique to organs as intimate as the penis, and the psychological impact of receiving an organ should not be underestimated. People wait anxiously for months hoping and praying that the phone will ring and they'll have won the organ lottery, but when they do it's not necessarily all peaches and cream. Anxiety, depression and fear of organ rejection are often experienced by people post-transplant. For some the emotional sequelae, the long-lasting conditions that can follow such an intervention, are more unusual, with a handful of people reporting changes in their personalities.

In an article in the *Journal of Near-Death Studies*, researchers reported a series of interviews with 10 heart transplant recipients, plus their friends and family. All the recipients reported changes in their preferences for a variety of things including food, music, sex and art. One woman described how her boyfriend had become a much better lover since having a woman's heart transplanted into his chest. His doctors suggested this was probably because he was no longer terminally ill, but the recipient asserted it was because he now had a woman's way of thinking about sex. A 29-year-old woman who received the heart of a 19-year-old who had died in a car crash explained that she could feel the accident that killed her donor: 'I can feel the impact in my chest. It slams into me.'

The power of preconception shines particularly strongly in one of the case studies where a white man (who asserted he was *not* a racist) thought that receiving an African-American heart might make his penis grow larger. Incidentally, his post-transplant taste in music took on a more classical leaning, but he dismissed the notion that this could be linked to his donor because he assumed his donor would have loved rap music. His donor was actually a violinist.

While this article would force even Fox Mulder to have a few sceptical thoughts, the point isn't whether a heart can transfer memories or enlarge a penis (hint: it can't). It's about the psychological impact of transplants on people's complex sense of self. It's also testimony to the powerful position the muscle-bound pump in our chest holds in our culture – we give our heart to the person we love, we encourage people to follow their hearts, we try to win hearts and minds. In a study exploring the impact of an organ donor's character on a person's willingness to receive an organ from them, people were least enthused by the idea of receiving a murderer's heart. Picture a murderer's heart – what colour is it? I'm willing to bet most people picture something black and twisted.

The cultural significance of our cardiac muscle is no doubt part of the reason why the world's first heart transplant caused a media storm quite unlike any surgical procedure before it. On 3 December 1967 in South Africa, Christiaan Barnard entered the history books for performing the first human-to-human heart transplant. Barnard has been described as both a charming playboy and someone prone to outbursts of temper who inspired intense dislike among some colleagues. However his drive and dedication were undeniable, and in preparation for the landmark operation he and his brother went as far as to perform heart transplants on dogs.

Their first human heart recipient was 53-year-old greengrocer Louis Washkansky, a diabetic smoker with legs swollen to Michelin-man proportions thanks to fluid build-up, and a heart that after three heart attacks had had enough. Barnard removed the tired organ and replaced it with the heart of a young woman who had been fatally injured after being hit by a car. Plumbing in the new heart was a surgical success, but sadly for Mr Washkansky it wasn't long lived. He died 18 days after the operation from pneumonia and septicaemia.

This didn't quell the enthusiasm of other centres across the globe and the world went from seeing a single heart transplant in 1966 to over 100 by 1968. However the success rates were woefully poor, with barely more than 50 per cent of people surviving a month after their procedure and just 11 per cent still alive after two years. This led many centres to abandon the procedure and only the Medical College of Virginia, Pitié-Salpêtrière University Hospital in Paris, Stanford

Hospital in California and Barnard's group in Cape Town continued to do heart transplants and advance the art. Now fast-forward to the present day and over 5,000 cardiac transplants occur every year with 85–90 per cent of patients living one year after their transplant and almost 75 per cent still alive after three years.

Rejection

While surgical techniques have improved, one of the biggest reasons for this transformation in transplant survival sits squarely in the realms of immunology. Central to the death of Louis Washkansky and the other early transplant recipients was medicine's limited understanding of transplant rejection and the drugs used to suppress it. Today we have a much better handle on why this happens and how to stop the immune system biting the (transplanted) hand that feeds it. In 1967, the year after the first heart transplant, the human major histocompatibility complex (MHC) was discovered. The MHC* is a chunk of DNA on the short arm of chromosome six that includes over 200 different genes, some of which code for markers called class I MHC molecules to be displayed on the surface of cells. These molecules act like barcodes that can be scanned to unveil whether the cell is a healthy 'self' cell. They are found on the surface of every single cell in the body that has a nucleus, from those on the tip of your tongue to the ones in the deepest, darkest recesses of your bowel. We inherit our collection of molecular markers as two sets, taking a 50 per cent selection from mum and a 50 per cent selection from dad to create our own full

---

* Fun nomenclature facts: the human MHC is also known as the Human Leukocyte Antigen (HLA). Some people use MHC to refer to the genes and the term Human Leukocyte Antigen for the proteins made by those genes; others use HLA and MHC interchangeably. The class I MHC molecules are also known as HLA-A, HLA-B and HLA-C, a fact I mention because we'll come across them again in Chapter 6.

complement of class I molecules. This selection approach
means we have a good chance of having an identical set of
class I MHC molecules to one of our siblings (which is why
they can be a useful source of spare parts).

The class I MHC molecules present mushed-up proteins
from the inside of the cell for the perusal of a white blood cell,
called a T cell. It's the T cell's job to spot infected or abnormal
cells. Stick in an organ from someone else and it's like putting
a pink polka-dot cushion on a zebra-print couch – the clash is
impossible for the T cell to miss. The body recognises that this
interloper is something new, something different, and attacks
it. The end result is that the T cells cause the organ to die cell
by cell, effectively triggering the transplant's suicide. This
process results in acute rejection and typically happens in the
first six months post-transplantation.

While acute rejection no doubt feels all too fast for the
recipient, it is glacial in comparison to hyperacute rejection,
whose antibody-led assault causes organ rejection within
minutes to hours of the transplant, as the blood vessels of the

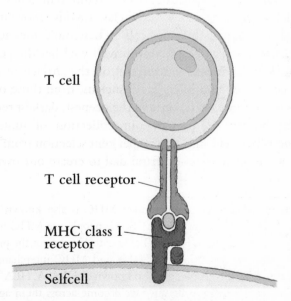

Figure 4.1 MHC class I presenting self antigen to a T cell.

organ are rapidly destroyed. Hyperacute rejection happens when the person receiving the transplant already has antibodies primed to attack the transplanted cells. This occurs when the recipient's immune system has previously been exposed to sufficiently similar donor cells, which can happen for a variety of reasons including previous transplants or blood transfusions. The antibodies rapidly attach to the transplanted organ, trip the complement cascade we met in Chapter 3 and block the blood vessels of the donor organ by ramming them full of dense blood clots. No blood flow > no oxygen > no living cells > dead transplant.

While the speed of hyperacute rejection is brutal, there's something almost worse about the slow, gnawing nature of chronic rejection. This pernicious phenomenon can rear its ugly head many months or even years after acute rejection has come and gone. A combination of T cell and antibody attacks on the organ inflicts progressive damage, scarring it slowly, steadily and unflinchingly. The way this manifests varies from organ to organ – so in heart transplants the scarring leads to accelerated hardening of the coronary arteries, whereas a transplanted liver develops vanishing bile duct syndrome.

Lung transplants suffer terribly from chronic rejection. It affects the majority of transplanted lungs and causes bronchiolitis obliterans syndrome (BOS). Any disease which sounds like 'obliterate' doesn't bode well, and BOS is no exception. It attacks the lungs, transforming them from vanishingly thin gas exchange machines to scarred, fibrous bags which oxygen and carbon dioxide have to battle through. BOS is a key reason why patients' median survival after a single lung transplant* is 4.6 years, which is shockingly low compared with the 15 years you'd expect from a kidney transplant. Sadly once BOS has set in, the deterioration in lung function is irreversible and unstoppable.

---

* Median survival means 50 per cent of people survive at least that long.

Although BOS cannot be reversed, in general being aware of the health of transplanted organs is an essential part of caring for these patients and mitigating the impact of rejection. However, this is no mean feat because in most cases it necessitates taking a tiny chunk out of the organ and looking at it down a microscope. In the case of heart transplants that involves taking a piece of the heart, which as you might imagine comes with a range of risks from causing an abnormal heart rhythm to perforating the heart. Not something you want to do at regular intervals, really. Yet as there's no alternative, heart transplant recipients have dozens of biopsies nicked from their hearts over time. Current research has been trying to change this by striving to find a non-invasive way of monitoring transplants and making the process as simple as a blood test. One promising avenue has been to look at the pattern of genes being expressed by patients' white blood cells, which as you might expect seems to vary significantly between those with a healthy transplant and those in full-blown rejection. AlloMap was the first example of a US FDA (Food and Drug Administration)-approved blood test to arise from this research and it's currently being used in clinical practice to spot heart transplant rejection. Unfortunately it has a high false positive rate, which means its results have to be interpreted with caution. However if viewed as one piece in a puzzle to be combined with other tests like ultrasounds of the heart and examinations by doctors, AlloMap can reduce the need for a heart biopsy without increasing the risk of serious cardiovascular events.

Another emerging option in rejection spotting is to look at the level of donor DNA in the recipient's blood. When cells in a transplanted organ are attacked by the immune system and die, they spew out their DNA, which can then be detected in very high levels in the recipient's blood, signalling the onset of rejection. Initial research in this field focused on gender-mismatched kidney transplants that involved a female receiving a male kidney. This allowed researchers to look for the male-only Y chromosome in the blood and urine of the female recipients. However this is clearly restricted to instances when women receive organs from men, which represents less than 25 per cent of all transplants. Scientists at Stanford University have developed an alternative test that detects donor DNA fingerprints in the blood of heart transplant recipients. It's early days and much work will need to be done before the test hits the wards, but preliminary reports (from the team making the test) suggest that it is safer, cheaper and more accurate than a heart biopsy.

Finding your perfect match

So, we are getting better at identifying rejection but prevention is better than cure, so what can we do to stop rejection occurring in the first place? Like finding any new life partner, the first trick is to find the perfect match. The second trick involves taking drugs (and therefore has less inherent scope for an appropriate dating analogy).

Finding the perfect match involves a checklist that looks as if Frankenstein's monster started an online-dating service. Factors can include blood type, tissue type, body size and the ability of the organ to stay fresh. This last fact is very important, with hearts and lungs lasting only four to six hours, while livers are good for 12 hours and kidneys have a shelf-life of up to 36 hours, which means they can put up with a longer time in transit and thus be sourced from further afield.

Organ size is particularly important for organs going into the chest, such as the heart and lungs, because there's a limited amount of space inside the rib cage. The abdomen isn't such a rigid box, and the extra wiggle room inside this spongy cavern creates more flexibility on the size of liver or other organ that can be squished inside it.

Blood type is obviously important for blood transfusions – flooding the body with the wrong type of blood can lead to itching, fever, chills and even death due to a massive immune system reaction to the foreign blood. However blood type also plays an important part in organ transplants including the liver, heart, kidneys and lungs. Alongside blood type, some organs are also tissue-typed using the class I MHC cell barcodes we met earlier. Matching these tissue types is done routinely for kidney and bone marrow transplants, where mismatching has been shown to have a significant impact on the long-term success of the transplant. In the case of kidneys this has led to some creative problem-solving to set up matching organ donation chains. Let's say Han Solo needs a kidney, but his girlfriend Leia, who is willing to donate one of her kidneys, is not a match. In a galaxy far, far (but less than 36 hours) away, Darth Vader needs a kidney but his good friend the Emperor is not a match. If Leia is a match for Darth (what are the odds?) and the Emperor is a match for Han Solo they could enter into a mutually beneficial swap. This is an example of a paired exchange donation. Another option is for the chain to be kick-started by a Good Samaritan donor, who has nobody in their life who needs a kidney but is willing to donate one simply because they think it is the right thing to do. This initial donation kick-starts many pairs like the Leia–Vader Emperor–Solo example to create one epically impressive chain of donation. At the time of writing, the longest kidney transplant chain in the US involves 68 people swapping 34 kidneys across 12 months and 9 states. This process is fuelled by the fact that Good Samaritan donations are a rare but incredible gift that can inspire people to keep on paying it forward even if their relative has already received their kidney. In an effort to encourage such generosity, living donors also receive points from the National Kidney Registry for their donation so if they themselves ever need a kidney they get a priority boost on the waiting list.

All you need is drugs

We've now explored the trials and tribulations of finding the perfect match, but even when all the stars seem aligned at the beginning of the recipient-donor relationship it can still end in rejection. For most transplants, the only way to achieve long-term donor-recipient harmony is heavy reliance on drugs to suppress the recipient's immune response.

The earliest attempts at immunosuppression were almost as lethal as the transplant itself. In the 1950s the X-ray marrow protocol was trialled in kidney transplant patients, which involved sub-lethal doses of total body X-rays followed by bone marrow infusions. The results were largely unsuccessful, with just one of 12 patients having kidney function that lasted more than 12 weeks. A variety of other immunosuppressant techniques were attempted in animal experiments including administering chemicals like toluene (a widely used solvent in paint thinners) and nitrogen mustard (a compound similar to mustard gas, a Schedule 1 substance in the Chemical Weapons Convention).

In 1959, the breakthrough the transplant community so desperately needed arrived thanks to one of these animal experiments. Treating rabbits with 6-mercaptopurine for two weeks stopped them from producing antibodies against a particular foreign antigen. This tolerance remained even after the drug was stopped. Crucially, the rabbits were still left with a functioning immune system as demonstrated by their ability to make antibodies to other foreign antigens (in the experiment they exposed them to a cow protein). A slightly less toxic derivative of 6-mercaptopurine was crafted called azathioprine, which was then tested in people. The first experiment was a disaster. The patient died within a month because the dose that worked in dogs was lethal to people. They halved the dose for the next patient, and for the first time in history doctors were able to stop the rejection process in its tracks. However it came at a high price: even at half the dose azathioprine wiped out the patient's white blood cells. They stopped the azathioprine to let the white cells recover and the organ rejection set in. Once the white cells recovered they re-started the azathioprine and the kidney started to thrive again but then the white cells dipped once more. He died within a month from sepsis. Over time clinicians began to understand how to balance the immunosuppression see-saw, and azathioprine is still widely used as an immunosuppressant drug in kidney transplants today.

As our knowledge of the immune system developed, so did our ability to manipulate it, which has led to an explosion in immunosuppressant drugs relative to the 1960s. Today we can block T cell activation with sirolimus, tacrolimus or cyclosporin, use mycophenalate to stop white blood cells dividing or push prednisolone to interfere with key immune messenger molecules. Unfortunately these drugs come with an intimidating list of potential side-effects, ranging from vomiting to bowel perforation and cancer. More recently, man-made antibodies have been designed to make safer immunosuppressant drugs that are just as potent as their predecessors but less toxic. For instance the lab-generated antibody daclizumab is a super-targeted immunosuppressant that blocks the action of IL-2, a chemical messenger that plays an important part in rejection. Its targeted nature means daclizumab also has a relatively low rate of serious side-effects, which makes using it less problematic. This is where the real power of the next generation immunosuppressants lies, because it means we can use them more readily and rely less upon perfect tissue matches. This allows organ matches to be made which would previously never have got past the first date.

So far we've discussed what happens when a recipient rejects their transplant, but sometimes rejection is a two-way street. Bone marrow and small bowel transplants bring with them a chunk of the donor's immune system that endows them with the capacity to reject their new bodies. This phenomenon is known as Graft Versus Host Disease (GVHD) and is caused by the donor T cells recognising that their environment is foreign and attacking it. The bigger the difference between the tissue type of the donor and recipient, the bigger the risk of GVHD. Normally the donor cells would be vastly outnumbered and it would be like a band of rebels taking on a vast army on its home turf, but in GVHD the recipient's own immune system is in a weakened state and unable to fight back. The symptoms can be as mild as a rash, or as severe as hepatitis, intestinal bleeding and coma. Unlikely as it sounds after listing those symptoms, sometimes a touch of GVHD can be a good thing. This is true for leukaemia patients receiving a bone marrow transplant because the combative donor T cells can kill off some of the remaining cancer cells. For patients where GVHD is causing more harm than good, the solution is immunosuppressant drugs. Doctors typically use the same drugs we met before – including tacrolimus, sirolimus and steroids – but this time given to reduce the risk of the donor rejecting the host.

Supply and demand

In the midst of this immunological warfare and arsenal of cutting-edge drugs, it's easy to forget that the biggest limiting factor on transplants boils down to a very simple equation: supply and demand. The US Organ Procurement and Transplantation Network currently lists over 123,000 people in the States who are in need of a transplant. In 2014 there were just 15,670 donors. Despite the fact that one donor can save eight lives, the US waiting list is growing at a rate of one person every 10 minutes. The result is that 21 people die waiting for a transplant every single day, and the gap between the number in need and the number donating is only getting worse.

Cue some organ supply-chain creativity. The year 2015 saw the announcement of Europe's first non-beating heart transplant. On the face of it, this might not sound entirely impressive – surely every heart stops beating before you swap it out? Yes, but it's all about the timing. Part of the reason the pool of heart donors is small is that people have to die in quite specific circumstances – they have to be brain-stem dead but still have a heart that's beating. It's a niche state. While someone whose heart has stopped beating can still donate their kidneys or liver, doctors normally worry that their heart will be damaged even by a brief period of this so-called circulatory death.

The operation in Cambridgeshire used a heart from a donor whose heart and lungs had stopped. They restarted the heart five minutes after it had stopped beating and monitored it for 50 minutes. Once the team were satisfied that the heart was strong and working well they carefully removed it from the donor's body and placed it in something that will be on every transplant surgeon's Christmas list: a heart-in-a-box. The heart snuggles in a clear box with tubes that ensure the organ receives a constant supply of nutrients and oxygen. The portable transfusion system also has lots of built-in monitoring to allow doctors to keep a close eye on the heart's health right up to the moment of transplantation. Unsurprisingly, the price tag for this piece of tech isn't cheap (and would have even the most generous Christmas elf doing a double-take): £150,000 ($196,000) for the machine and £25,000 ($33,000) per transplant. However, the Cambridgeshire transplant surgeons who used the technique believe it has the potential to increase the number of hearts available by up to 25 per cent. It's not easy to put a price on that.

Another way of increasing the organ supply is to make the ones we do have go further. Take the liver: this hefty wedge-shaped organ weighs about 1.4kg (3lb) in an adult. As early as 1984 surgeons began making the most of this fact by using one adult liver to provide two transplants: one for an adult and one for a child. Over the last 30 years the rate of children dying on the waiting list for a liver transplant has plummeted, thanks to this technique. Not only has this surgical technique made better use of deceased donor organs, it has also enabled live donor transplants with parents donating a small segment of their liver to their sick child.

A third, and sometimes more controversial, option for increasing the donor pool is to look beyond our own species. The origins of xenotransplantation (transplanting non-human tissue into humans) actually extend further back in time than human-to-human transplantation. In 1906 a surgeon by the name of Jaboulay had a bizarre kidney zoo, attempting transplants from goats, monkeys and sheep into people (with a 0 per cent success rate). The success rate didn't improve over time and consequently interest in xenotransplantation subsided. However it was re-ignited in the 1970s when renal dialysis took off and created further demand for organs by increasing the number of people being kept alive with incurable end-stage renal failure.

This revival was stopped dead by a pig virus. The discovery of porcine retrovirus in the 1990s had xenotransplantation researchers downing tools as clinical trials in the area were banned, so that the risks of cross-species infection could be investigated. The fact that it was a pig virus was particularly unfortunate because pigs are seen by many as the species of choice for animal-to-human transplantation. This is because they are plentiful, have many progeny and handily have organs roughly the same size as ours (so no clever carving of the organs is called for). Moreover as they are already part of our food chain the general public have fewer ethical qualms about using pigs for the benefit of mankind. This contrasts sharply with endangered primates like chimpanzees, who were the original poster child for xenotransplantation.

However there is one massive barrier to non-primate xenotransplantation: the feisty human immune system. Our defence system is able to pick up subtle differences in tissue type between related people, so you can imagine what happens when it finds a hunk of pig has made itself at home. Hyperacute rejection kicks in within a few minutes to hours and the pig cells are annihilated by an antibody-led attack. The reason the rejection is so rapid is that we already have pre-existing antibodies known as xenoreactive natural antibodies (XNAs) circulating in our blood, which can bind to non-primate cells and tag them for destruction. You may, quite rightly, be wondering how we can have pre-formed antibodies to pig cells if we've never been exposed to them before. The answer lies inside our own gut. The healthy human gut is full of bacteria, many of which have a carbohydrate marker on their surface that isn't found in people. This same carbohydrate marker is found on the surface of pig cells, so the XNAs we develop in infancy to attack the bacteria in our gut are also able to attack transplanted pig cells.

In an attempt to overcome this problem scientists created genetically engineered knockout pigs,* which had the gene responsible for the carbohydrate marker erased from their DNA. The researchers then tried transplanting these altered pig hearts into baboons, and found that while they overcame the hyperacute rejection problem, the hearts still only lasted six months. A more advanced form of disguise was required. So the next stage in disguising the pig cells has been not only to remove the tell-tale carbohydrate marker that screamed 'pig cell' but also to add in genes for human markers. A bit like handing the pig cell a fake moustache and glasses to confuse the immune system.

These undercover pig cells have yet to be trialled in humans and it will be some time before major organ xenotransplant trials become a reality. However there is one form of pig-based transplant that made it into people over 50 years ago and is still in use today: pig heart valves. Roughly 280,000 heart valve replacements are fitted worldwide every year. About 50 per cent of these are bioprosthetic valves made either from a pig's heart valves or from the tissue sac that surrounds a cow's heart. Thus pig heart valves are part of mainstream medicine, and are sometimes the preferred option to metallic valves because they don't require a lifetime of blood-thinning drugs, which are a must with metal valves. They also have reasonable longevity, with the vast majority of bioprosthetic valves lasting for over a decade because they don't stimulate the immune system as strongly as a whole organ transplant like a kidney or liver.

In part this is because the pig heart valves are treated with a pungent oily chemical called glutaraldehyde, which is also used for industrial water treatment and the sterilisation of medical equipment. The glutaraldehyde alters the structure of collagen in the valves and in doing so masks the markers

---

* Knockout animals have been genetically modified to inactivate, or knock out, an existing gene. Knockout mice have been used in a vast array of trials investigating everything from obesity to cancer.

that betray the valve's porcine origins. This dramatically reduces the risk of rejection, but it also damages the cells, leading to an increase in the amount of calcium in them. Over time the valves become calcified and eventually fail due to tearing or stenosis (narrowing of the valve). The glutaraldehyde technique isn't viable in something as large and complex as a whole organ, so it is restricted to heart valves for now. Moreover it isn't perfect when it comes to hiding the pig cell antigens, and it seems that a low-grade immune response triggered by the few remaining pig antigens grinds away at the graft slowly but surely over years.

On the plus side, the glutaraldehyde does seem to tackle the pig virus problem, which remains a major hurdle for transplanting solid organs from pigs to people. In fact the problem isn't limited to pigs. A heady cocktail of viruses that have never encountered people, mixed with immuno-suppression, means that transplants from other species carry an unavoidable risk of infection that exceeds that of human-to-human transplants. A virus can be relatively innocuous in its normal host, but can become deadly in a foreign species. For instance, B virus looks a lot like herpes when it infects its normal host, the macaque monkey. However, if it gets into people, B virus causes an infection in the brain that kills 70 per cent of those who catch it. This unpredictability makes xenosis (infections spread through xenotransplants) a fear-inspiring spectre that haunts the field of xenotransplantation and limits its application.

The situation isn't helped by the fact that some of the most dangerous infections on the planet, including HIV and the Ebola virus, began as animal diseases and then leapt the species barrier into people (though not via xenotransplants). These concerns mean that any attempt to transplant an organ from an animal into a human will come with a programme of monitoring and surveillance intense enough to make Edward Snowden sit up and take notice. While it makes perfect sense, this approach raises certain ethical concerns, because it effectively means that once someone receives a xenotransplant they lose their right to exit the experiment.

The right to drop out of a clinical trial at any point for any reason is enshrined in the Declaration of Helsinki, the bible of ethics for clinical trials, so contravening this is a massive consideration for researchers in the field and for anyone considering a xenotransplant. While big trials of major organ xenotransplants are still a long way off, these sorts of ethical issues need to be considered in depth long before anyone picks up a scalpel or signs a consent form.

In my humble, and painful, experience, medical ethics is a subject that often seems drier than the dusty remains of Tutankhamun. However, when it comes to transplants the importance of ethics should never be underestimated. This can best be demonstrated with three simple stories: the path from death row to donation, the street value of a pound of flesh and the tale of Baby Fae's broken heart.

From death row to donation

Fact 1: China executes more prisoners than any other country in the world. Fact 2: like every other country in the world China has a significant supply–demand problem when it comes to transplants. China combined these two facts and came up with an organ supply chain that connected death row and organ donation. A violent twist to this already dark tale is that prior to 1 January 2015, the organs of executed prisoners were taken without consent and distributed not necessarily to those most in need but (illegally) to those with the deepest pockets. Timings of executions were allegedly optimised so that organs could be made freshly available for well-connected, wealthy recipients. However, despite all of this the number of organs available still fell far short of meeting the demands of the Chinese population and the approach caused a significant outcry from the international community. Thus since 1 January 2015 the organs of executed inmates should no longer be taken for transplants. Instead an organ bank has been set up to receive donations from deceased volunteers across China. The head of the government's organ donation committee hoped that by the end of 2015, 10,000 organs would be donated, matching the number of transplants typically carried out in China each year. That's still a long way short of the number needed by a country the size of China. In 2011, one million people in China were in need of a kidney, according to their national kidney transplant registry, but just 5,253 actually received one. That leaves 99.5 per cent of those on the waiting list with not a lot of hope. With stats like that it's not hard to understand why some people suggest the ethically abominable, but highly lucrative, black market in death row donors will be hard to stop.

The street value of a pound of flesh

Organs from prisoners come with a higher risk of blood-borne viruses like hepatitis C and HIV. Which is why some people willing to buy an organ with questionable provenance decide to request a free-range one. Thanks to these recipients the black market in organs is alive and well, with a kidney fetching up to £128,000 ($167,000) in countries like China, India and Pakistan. This illegal market typically harvests organs from the poor and the vulnerable, who receive a tiny fraction of their kidney's price tag with much of the profit going to doctors, agents and various middlemen who have risked neither life nor limb. It's been estimated that 1 in every 10 transplants that takes place worldwide involves illegally trafficked and purchased human organs. There have even been reported cases of people being murdered simply so that their organs can be harvested and sold.

While we can all agree there's no ethical ambiguity about murdering people for organs, some ethicists have suggested that people should be allowed to decide to sell their own. They argue that if we had a legitimate market for organs we could increase the supply and allow people to make an informed choice to sell their organs at a fair price in a regulated market. Iran is the only country in the world to have made the sale of kidneys legal. In fact there is stiff competition among would-be donors who stick up homemade adverts on the streets approaching Tehran's major hospitals. Two non-profit organisations monitored by the state are charged with finding potential sellers and linking them up with recipients in a regulated system. Critics point to the fact that donors are disproportionately made up of Iran's poor and that little follow-up has been conducted to understand the long-term health of these donors. However, the outcomes of recipients have been good, and unlike the UK, the US or China, Iran has had zero people on a waiting list for a kidney since 1999.

Baby Fae's broken heart

In 1984, Baby Fae was born with a broken heart. There are conflicting accounts of her first few days, but it appears her parents were told she had no hope of surviving, that she would sleep for longer and longer periods and eventually pass away. They took her to a motel to say their goodbyes and wait for their little girl to die. However, shortly after leaving the hospital they got a phone call offering a massive gamble with a tiny hint of hope: experimental surgery to replace her poorly formed heart with one from a baboon. They agreed, and a senior surgeon from the hospital, Dr Leonard L. Bailey, cut his holiday short and hurriedly set about making arrangements for the experimental surgery. Approval to attempt the surgery had only been received one week previously, and the revelation that Bailey didn't even ask organ procurement agencies to look for a human heart left questions over whether Baby Fae's survival, or the desire to experiment, drove the decision-making. Reports at the time said Dr Bailey told the mother that her baby would experience three episodes of rejection, something that was surely hard to predict given the limited precedent. He also told *American Medical News* 'We're optimistic that within three months, she'll be able to go home.' This seems more than optimistic given that only three people prior to Baby Fae had received a heart from a chimpanzee or a baboon, and their survival time ranged from 90 minutes to three and a half days. This ardent optimism, combined with shifting reports from the hospital as to how frequently consent forms were signed by the parents, raised many questions about just how well informed their consent for the procedure was. While she held on for longer than any patient before her, Baby Fae's body eventually rejected the baboon heart. Twenty short days after the walnut-sized heart was sown into her chest it gave up. Brief as her little life was, Baby Fae's story caused a stir that still ripples through discussions of xenotransplantation and the ethics of informed consent in experimental treatments today.

The cutting edge of transplantation

These three stories highlight the fact that immunology isn't necessarily the most complex challenge in the field of transplants. This is most apparent at the cutting edge of transplants where doctors, scientists, patients and the public are all trying to tiptoe through an ethical and technical minefield. The potential benefits for being willing to walk this line can be immense. In February 2015 the *Lancet* reported a true miracle baby – a little boy who grew inside a family friend's uterus that had been transplanted into his mum. As with the vagina transplant recipients we met at the beginning of this chapter, the uterus transplant recipient had MRKH syndrome, which had left her without a uterus of her own. Her post-menopausal friend no longer needed her own womb, so she donated it, and one year after the transplant the 36-year-old recipient underwent IVF and became pregnant first time. It wasn't all plain sailing and despite taking powerful immunosuppressant drugs before and during the pregnancy, she suffered three mild episodes of rejection. Thankfully these episodes of rejection didn't cause her to lose the baby, but at just shy of 32 weeks she developed pre-eclampsia and had a caesarean section. Happily the baby was a very healthy boy who was a normal weight for his age. There is no bigger stress test imaginable for a transplanted organ than pregnancy, so this was a landmark achievement. Yet despite this happy ending there are lots of challenging issues raised by this case. For instance, is it ethical to have two people undergo major surgery, to have someone exposed to powerful immunosuppressant drugs, including during pregnancy, not to save a life or even to bring a child into the world (surrogacy would do that) but primarily so someone can experience pregnancy? Perhaps if everyone involved has consented to the plan and knows the risks it's their call? Perhaps.

A working uterus is one form of extreme transplant; however, our next example of the cutting edge of transplants is less about the organ in question and more about scale. So far we've spoken about transplanting one kidney, one liver, maybe a pair of lungs, but how many transplants can you do at once? In 2011, nine-year-old Alannah Shevenell had six of her organs replaced in one incredible operation. Alannah had an enormous tumour growing in her abdomen, which doctors had tried and failed to remove on two previous occasions. On the third successful attempt they removed not only the tumour but also her stomach, liver, spleen, small intestine, pancreas and part of her oesophagus. These were all replaced with transplanted organs. This mammoth operation was a massive undertaking not only for the surgeons, who had never done anything like it, but also for little Alannah. Her doctors at Children's Hospital Boston advised Alannah's family that her chances of surviving the operation were only 50:50. Demonstrating the strength of the human body and mind in bucket-loads, Alannah came through it and continues to do well, though she will need to take immunosuppressant drugs for life.

As doctors, patients and researchers push science forward the lines between fantasy and reality are becoming blurred. Take *Face/Off*, a 1997 film in which police officer John Travolta swaps faces with criminal Nicholas Cage in a bid to prevent a terrorist attack. In 1997 this was pure fiction but just eight years later surgeons conducted the world's first partial face transplant. The recipient was 38-year-old Isabelle Dinoire, who had attempted suicide by taking an overdose and passed out. She came around after her Labrador retriever, desperate to wake her, bit at her face, removing her entire chin, the end of her nose, her upper and lower lips and parts of both cheeks. After 15 hours of surgery she woke to find she had a new face to get to know, a patchwork of her own sewn together with someone else's. The results of the operation were far from the perfection of *Face/Off*, but Isabelle was pleased to have a mouth and a nose to be able to eat, speak and breathe through. However, she also awoke with a notoriety she couldn't have dreamt of and was pursued relentlessly by the press who wanted to capture a photo of the woman with someone else's face. Over time she came to learn to live with her newfound face and her newfound fame: 'When I feel down, or depressed, I look at myself in the mirror… She gives me hope.' Sadly Isabelle died in April 2016 at the age of 49 from cancer thought to be the result of the profound immunosuppression necessary to stop her body rejecting the face transplant.

With this statement Isabelle gives voice to an idea that has echoed throughout our story of transplantation. Every pound of flesh we're clad in and every organ jammed inside us, even if we don't give a second thought to it or frankly know its name or precise location, plays a primal, undeniable role in our definition of self. Not simply in the immunological sense but in a deeply personal and philosophical 'who am I?' sense. Having a part of another person joined on to you, however seamlessly, can challenge this sense of self – for better or worse.

But maybe the story is even more complex than that. Maybe it's not just me or you, maybe it's me, you and *them*. You see, there is another set of characters skulking around our innards which may influence our everyday lives: the billions of bacteria that line our gut and live on our skin making up our microbiota. Emerging evidence suggests they may have the capacity to affect not only the health of our bodies but the contents of our thoughts. Let's meet our tiny masters.

CHAPTER FIVE

Outnumbered: How the Immune System Manages the Body's Resident Bacteria

My friend Zain works at a *very* special bank. People come to Zain's bank to make transactions – money exchanges hands, the staff are courteous and professional and the customers leave satisfied. Excellent customer service aside, it doesn't sound too unusual for a bank, until you realise the nature of the customers' deposits: their own poo. These customers are actually registered poo donors who come to add their contribution to the growing field of poo transplants. The bank is OpenBiome, a Boston-based non-profit organisation which started at the world-class research institute MIT. Their aim is to improve access to safe faecal microbiota transplants, which is a more genteel and scientifically accurate name for poo transplants.

You see it's not the poo that Zain and his colleagues care about – it's the medical gold it contains, namely all the bacteria that make up the faecal microbiota. Not all of our microbes can be found in our poo and the human microbiota is the collective name for the 100 trillion micro-organisms that have made us their real estate. From the tip of your tongue to the skin you sit on, they have set up home in every intimate nook and cranny of your body, and as we'll find out in this chapter, that's largely a fact to be grateful for. Indeed our own cells are outnumbered by these resident aliens, with some estimates suggesting each of us carry around 10 times more micro-organisms than human cells. Thankfully we don't all look like a writhing mass of microbes because the cells of the microbiota are much smaller than ours, so they constitute just 1 to 3 per cent of our body mass.*

The man-versus-microbe figures are even more staggering if you look at the number of microbial genes we're walking around with. This exotic collection of foreign genes is called the microbiome (although some people use the terms microbiome and microbiota interchangeably). Over time, the more we've learned about the human genome, the smaller the number of active genes has become. In 1990 the US government estimated there were 100,000 genes but as methods to count genes accurately came into play the number tumbled and at last count there were thought to be 22,000 active protein-coding genes in the human genome. That's more than a fruit fly, but less than *Trichomonas vaginalis*, a single-celled parasitic organism that causes roughly 180 million urogenital tract infections in people every year. This pest to our privates has a stunning 60,000 protein-coding genes at its disposal, almost three times the number of active

---

* This still means most of us can blame at least one or two pounds of our weight on our microbiota – a fact to squirrel away for post-Christmas/Thanksgiving/general indulgence encounters with the weighing scales.

genes necessary to run an entire person. The story of the shrinking human genome contrasts sharply with the human microbiome, where the more we've learned the bigger the numbers have become. The most recent cataloguing reported 3.3 million microbial genes in the human gut alone, which means microbial genes outnumber human genes by a factor of more than 100:1.

It's the actions of these genes that makes our relationship with these microbes so fascinating: from synthesising vitamins to making potentially mood-altering neurotransmitters, many interact directly with our physiology. As well as providing more benefits than a box full of superfoods, there is also massive diversity between our inner menageries. If you consider the human genome, we are all about 99.9 per cent identical, but when you look at our microbiomes you find we are as much as 80–90 per cent *different*. This variety is why some scientists predict it's the Human Microbiome Project, not the Human Genome Project, which will harbour the most exciting scientific discoveries in terms of personalised medicine in the coming years. From bespoke probiotic yoghurts for mental health problems to bacteria transplants for acne, this is a field with an abundance of material for both cynics and optimists to revel in.

The prime real estate for these microbes, the Manhattan or Mayfair equivalent inside you and me, is the large intestine or colon. If you had a Lonely Planet or Rough Guide to your gut, the colon would have an entry something like this: 'The colon is a must-see multi-cultural melting-pot, where up to one thousand species of bacteria mingle and dine together every second of every day. In this truly 24/7 subterranean city, *Enterococci* rub shoulders with *Clostridia*; *Bacteroides* luxuriate in their oxygen-depleted surroundings and *Bifidobacteria* banquet on a sumptuous all-you-can-eat poo buffet. It's the microbe's place to see, and be seen.'

However, sometimes the happy balance of this poopy paradise can be disturbed. Among the key disruptors are doctors, because they prescribe antibiotics. Antibiotics are undeniably one of the most important discoveries in medicine;

since the Second World War they have been transforming the course of human history by making the terminal survivable. However, they can have side-effects, and one example is their potential to kill off vast swathes of the normal gut flora. This creates an open-plan living space for a hardy bacterium called *Clostridium difficile*. This so-called superbug (also known as *C. diff*) is able to survive the initial antibiotic onslaught and then rapidly multiplies in its newly vacated palace. The result for the patient varies from watery diarrhoea and stomach cramps, to torrential diarrhoea, perforation of the colon and death. In the US, *C. diff* is the most common hospital-associated infection, causing an estimated 453, 000 infections in acute hospitals in 2011 and costing the nation over $4.8 billion annually in the process.

The mainstay of treatment to date has been even more hardcore antibiotics chosen for their ability to destroy *C. diff*. However, they don't work for everyone. So, cue a bit of blue-skies thinking. If the problem is that the normal microbial residents have gone, why don't we just pop them back up there to crowd out the *C. diff*? What's more, unlike antibiotics, we have a free and never-ending supply of these bacteria, which we're currently flushing down the toilet. That's the essence of a poo transplant, which transfers the faecal microbiota of a healthy volunteer into the colon of someone suffering from *C. diff* through the medium of medicinal poo. These healthy volunteers are the intestinal elite as only 2.8 per cent of applicants to OpenBiome are successfully selected, which means it's easier to get into Harvard than become one of their poo donors.

In the early days of the treatment, it was a fairly rustic process. Donor samples would be tested for common diseases and the donor would be advised to avoid eating anything the recipient was allergic to. The whole donation would be blitzed in a regular household blender with some salt water until thoroughly mixed. The slurry would then be strained through a coffee filter to remove larger matter and brew a microbe-rich consommé. Next came the logistical challenge of getting people to consume it. Thankfully this step was less rustic and it could either be delivered directly into the patient's small intestine via a tube passed down through the nose or alternatively up into the colon via an enema or colonoscopy. While it sounds like a scene from the most horrific episode of *Ramsay's Kitchen Nightmares* ever, this repulsive recipe saves lives. In the first (albeit small) randomised controlled trial of faecal transplants, published in the *New England Journal of Medicine*, the technique cured 94 per cent of recurrent *C. diff* infections. In fact this was so much better than the 23–31 per cent cure rate in the groups treated with antibiotics, that the trial was ended early because it was unethical NOT to give those groups faecal transplants too.

It may sound unpleasant, but for people with recurrent *C. diff* infections this rich brown elixir promises freedom from the exhaustion and social isolation of crippling bouts of severe diarrhoea. Some even attempt DIY faecal transplants using a donation from a loving and open-minded spouse using instructions from the internet (there are YouTube videos, though not for the faint-hearted viewer). Organisations like OpenBiome have refined the process, and exist to make it as safe and acceptable to recipients as possible. Prospective donors complete a medical questionnaire and undergo blood and stool screening. They agree to make regular deposits for a 60-day period, and can earn up to $250 a week. The poo is screened, prepared and frozen all ready to be shipped to waiting clinicians to administer it to patients. Just 50g (1.8oz) of poo can transform someone's life.* What's more, for any medical managers the poo transplant is like something out of a pharmaceutical fairy tale: a treatment which is cheap, plentiful and unlikely to run out anytime soon. There is of course demand for a synthetic version, the hope being that scientists could create lab-grown strains of bacteria in a stool substitute that provides all the therapeutic benefit of poo transplants without their 'ick' factor. Trials are currently

---

* So for anyone who wants to help their fellow man, without losing a vital organ in the process, literally all you need to do is give a … poo.

ongoing, but if 'Repoopulate's' scientists are even half as inventive as the committee that named the product, there are exciting times ahead.

Curing crippling diarrhoea isn't the only potential perk from our relationship with our gut bacteria. In 2014 Bahtiyar Yilmaz and colleagues showcased evidence that certain gut bacteria can trigger a protective immune response to malaria. Both the malarial parasite and a harmless gut bacterium called *E. coli O86:B7* display a surface marker called alpha-gal. People who count this bacterium among their normal gut residents naturally create antibodies that recognise the alpha-gal, allowing the immune system to keep the *E. coli* strain in check. Yilmaz and colleagues suggest the antibodies have a second benefit in that they continue to circulate in the bloodstream and are therefore available to give a hostile welcome to alpha-gal-wearing malaria parasites. The study found two key bits of evidence to support this premise. First, people with high levels of alpha-gal antibodies are less likely to develop malaria, and second, mice which have been given alpha-gal-attacking antibodies are protected against catching malaria. So it seems our gut bacteria effectively provide a training ground for the immune system – a boot camp led by billions of bacteria which teach us to develop an arsenal of antibodies to tackle common foreign invader fingerprints like alpha-gal.

The flora of your face

Yet before we start to feel too warm and fuzzy about our resident microbes, I have two thoughts for you: face mites and acne. Crawling across your face right now, at a rate of 8–16mm/hour (0.3–0.6in/hour), are microscopic mites called *Demodex*. These semi-transparent critters are about 0.3mm (0.01in) long and are shaped like a chubby worm that's been lightly squashed, with eight stumpy legs crammed up around its head and a mouth that looks like it's full of pins. They come from the arachnid family (yes, like spiders) and like to nestle down in the hair follicles at the roots of our eyelashes, coming out to play at night. For most of us, having these incy wincy arachnids on our faces isn't a disease; they are actually regarded as part of the normal skin microbiota. Although occasionally, for reasons that are poorly understood, people develop Demodicosis, leading to itchy eyes and moulting eyelashes.

The *Demodex* dine on sebum (the waxy secretion we make to help waterproof our skin), as well as occasionally munching on our skin cells and even some unlucky commensal bacteria[*] like *Propionibacterium acnes*. These bacteria bring us to our second delectable topic: acne. This common condition affects 40–50 million people in the United States at a cost of $3 billion annually. About 80 per cent of people on the planet will experience acne at some point in their life, making it a significant global cause of both painful skin and psychological distress. Despite the scale of the problem, and the optimistic promises of the beautiful (and maddeningly acne-free) models on cleanser ads, acne is not a problem the dermatological community has solved. Scientists trying to pick at the problem repeatedly turn to the role of the many bacteria which reside on our skin. One of the main avenues of investigation focuses on the role of *P. acnes*. This bacterium is a frequent feature of the normal skin microbiota and, like many of us, *P. acnes* is a lipophile, which is to say it adores consuming fat. The sebum on our skin is like a layer of buttery, greasy goodness that has *P. acnes* smacking its lips. However, when *P. acnes* turns up to dine it has some seriously bad table manners, which can include dribbling chemicals over our faces. These chemicals can attract neutrophils and cause inflammatory symptoms like pain, redness and swelling, which are typical features of acne. Moreover, carbohydrates in the cell wall of *P. acnes* are also capable of triggering the complement cascade, causing further spirals of inflammation which contribute to the familiar angry red appearance of acne. It takes weeks for the

---

[*] The commensals are the micro-organisms that ordinarily live on and in the body.

body to degrade *P. acnes*, and this stubborn persistence means acne is often a chronic nuisance that takes time to clear.

So if *P. acnes* is the problem, surely antibiotics to wipe it out are the solution? Certainly antibiotics have been at the heart of acne treatment for over 30 years, but there's a blemish in this logic because *P. acnes* is the major bacterial commensal in both healthy *and* acne-affected skin. So why do some people with *P. acnes* develop acne, and yet those in the cleanser ads have the sort of skin normally only achieved by the airbrushed or the embalmed?

Decades of research have failed to figure out the acne conundrum, in part because mice don't get zits. In fact, there is no animal model for studying acne because animal sebum lacks the triglyceride fats that *P. acnes* loves to picnic on. Adding to the woes of the acne investigator is the troublesome fact that only a minority of microbes can be grown in lab conditions. This annoying problem hampers research into our skin's microbial multitudes. Most wither and die as soon as they are removed from the skin, making it impossible to study them. However, this all changed with the advent of next-generation DNA sequencing techniques, which have allowed us to identify the species and strain of bacteria rapidly, precisely and relatively cheaply, without ever needing to grow them in the lab. Scientists around the world are using this technique to build microbiome reference libraries to produce an accurate index of the many residents of our bodies based on their tell-tale DNA sequences.

A 2013 study took advantage of microbiome sequencing to try and solve the *P. acnes* puzzle. They compared the *P. acnes* of 49 acne sufferers with the *P. acnes* of 52 people with healthy skin, and found evidence to suggest that not all *P. acnes* are equal. The study showed two strains of *P. acnes* called RT4 and RT5 were more common in people with acne, while a third, called RT6, was found in abundance on healthy skin. The authors highlighted the fact that many other factors could be at play, from sebum production to hormone levels, and recommended that larger studies in future explore this relationship. However, the strong association between certain

P. acnes strains and skin type has thrown open a hitherto unexplored avenue for treating acne – one which targets the balance of the types of *P. acnes* instead of wiping them all out. Face-bacteria transplant anyone?

Making our menagerie

One thing demonstrated by the complex stories of the microbiota and disease is that when it comes to defending our body, the immune system isn't working in isolation. When fighting on certain fronts, such as preventing *C. diff* taking hold, it has the assistance of trillions of microbes that make us their home. However, managing these miniature mercenaries is no small task – after all they exist to satisfy their own interests, primarily getting fed. Microbes aren't easy bedfellows for a system whose *raison d'être* is to destroy foreign invaders and, as acne demonstrates, our inflammatory response to the microbiota can cause problems itself.

So how did this edgy alliance come about, and how can we influence it? Well, it's as old as we are and since day one its story has been told through the colourful medium of poo, beginning with baby's first poo, known as meconium – a tarry olive-green gift to new parents. Meconium is made of a mash-up of mucus, the baby's own downy hair, intestinal cells and secretions. Medical students (myself included) used to be taught that in some ways this was the pinnacle of our poo-producing skills because babies crafted something unique: a microbe-free turd. However in recent years scientists have come to question the idea that meconium is the cleanest poo we'll ever produce. A small study looking at the meconium of 52 babies born between 23 and 41 weeks found that over half had evidence of bacteria setting up shop in their guts, with bacteria more commonly found in babies born before 33 weeks. They also found that the gestational age of the baby when it was born had a significant effect on the types of bacteria in the meconium, with premature babies more likely to have bacteria normally associated with generating inflammatory responses. This was a small study that generated a lot of interesting questions, for instance, where did the bacteria come from? Did the type of bacteria play a part in causing premature labour? Unfortunately the study couldn't conclusively answer these questions. Instead it lobbed a few more pieces into a puzzle box that is already full of bits of various shapes and sizes, some of which aren't even part of this jigsaw. Also there's no picture of what the completed puzzle should look like and we've no idea how many pieces we should expect to find.

Yet there's hope in the impossible mission because some scientists raking through the box of the microbiome jigsaw puzzle think they've fished out some edge pieces. For instance, multiple studies suggest our method of delivery into the world has a significant impact on what our microbe menagerie looks like. Babies delivered via the vagina get a thorough coating of mum's vaginal residents, and within 20 minutes of birth the baby's microbiota looks a lot like mum's vaginal occupants. This contrasts with babies born by a caesarean section, whose microbiota more closely resembles that of their mum and dad's skin. These microbes seem to be an important part of the training ground for our immune system – just as siblings teach us social skills, so our inner zoo shapes the behaviour of our immune system. It follows that the mix of microbes we have could therefore play a part in our susceptibility to disease. Following this logic, some people have questioned whether the difference in the microbiome community in caesarean section babies could explain why babies born that way are at higher risk of asthma, obesity and type-1 diabetes. In fact one acclaimed researcher in the field was so concerned about the impact of a C-section on his own newborn baby's microbiota that he and his wife decided to bring some vaginal microbes to the party by coating a sterile swab in his wife's vaginal secretions and gently stroking them over their newborn baby's skin. It's unclear what impact this had on their daughter's gut flora, but it's a perfect example of just how quickly parents can become embarrassing.

While the microbe zoo we develop as an infant is important, it is by no means our definitive collection. Instead our microbe menagerie shifts over time, and can be influenced by many more factors than the stage entrance we use into the world. Some of the influences are quite peculiar, such as the association Ding and Schloss described in their 2014 *Nature* paper between educational background and vaginal bacteria. They found women with a baccalaureate degree were more likely to have a vagina teeming with *Lactobacillus type E*, while those without the baccalaureate were more likely to have low levels of *Lactobacillus* and moderate levels of *Atopobium*, *Prevotella*, *Bifidobacterium* and *Firmicutes*. Nobody mentions *that* in their university prospectus.

Trying to explain the connection between vaginal microbes and an undergraduate degree is a bit like trying to explain all 357 episodes of *Dallas* succinctly – it's simply too complex and bemusing. However, other connections are more straightforward, such as diet and the microbiota. The role of diet in shaping the microbiome starts early, with breast milk triggering the first seismic shift in the composition of our gut bacteria. The discovery of this shift actually solved a long-lasting breast milk mystery, involving a sugar without a purpose. Oligosaccharides are sugars found in breast milk that aren't readily digestible by a baby's gut cells, which has never made a lot of sense because evolution's pinnacle of foodie perfection shouldn't include indigestibles. It would be a complete waste of mum's mammaries to make a pointless sugar. However, the mystery oligosaccharides can be consumed by *Bifidobacterium infantis*, a normal resident of the healthy infant gut. By providing an all-you-can-eat oligosaccharide buffet for *B. infantis*, the breast milk encourages the bacteria to proliferate rapidly and spread like a benign blanket across the surface of the gut, leaving little room for other potentially harmful bacteria to grow. It also seems that the bacteria release chemicals that help shore up the intestinal defences by reinforcing the tight junctions which link the gut cells together. By aiding in the repair and maintenance of these linking points in our gut's security fence, our little caretakers help protect the integrity of the gut as a physical barrier. To reap these rewards babies not only feed the bacteria with breast milk, but they also appear to roll out the red carpet by suppressing their immune system, which would normally take exception to letting a gut-load of bacteria set up shop in the body.

We continue collecting various microbes throughout the first few months and years of life, like a bacterial version of baseball cards or Pokémon Go. Yet even in adulthood we never catch them all, because the types of bacteria we ingest and which thrive inside us are massively influenced by the types of food our environment provides for us.

In a quest to explore the impact of different diets on the microbiota, a group of researchers led by anthropologist Cecil Lewis took a journey deep into the bowels of the Peruvian Amazon rainforest. Almost literally. They sought out the Matses tribe, some of the last humans on the planet who are true hunter-gatherers, eating tubers that grow in the forest and hunting alligator, sloth, monkeys and more. One of the team's first challenges on getting out of the canoe and greeting their potential study participants was to explain to people unaware of the concept of the microbiome, why they should hand some of their poop to a bunch of random strangers who'd really, really like it.

Step one in this consent process was to explain they were interested in tiny organisms living inside the Matses' guts. To do this they brought a microscope and let the Matses see the microbes for themselves. Having convinced the Matses they weren't deranged, the team were rewarded with poo samples from 25 members of the tribe. They then went to the Andean highlands and made the same odd request of the Tunapuco people, who are potato farmers whose diet includes guinea pig, pork, lamb and cheese.* Perhaps deciding it wouldn't be polite to ask people to do something you wouldn't do, the

---

* Guinea pig isn't as unusual an appetiser as you might think and is found widely on menus in Peru. I found this out in a restaurant in the Peruvian capital, Lima, where the people on the table opposite nibbled on a grinning roast guinea pig wearing a jaunty upside-down half-tomato as a hat.

third group of people the researchers approached were academics from Norman, Oklahoma. Their diets were typical of WEIRD people (Western, Educated, Industrialised, Rich and Democratic) – think processed foods and ready-meals. Having gathered their samples of faecal gold, the team used cutting-edge gene-sequencing techniques to search for bacterial DNA signatures to reveal which bacteria lived in the participants' guts.

They discovered that the Matses and the Tunapuco had a much more diverse community of inner residents than the WEIRDs of Norman. In particular, they found several strains of spiral-shaped bacteria called *Treponema*. The *Treponema* have a varied and colourful family tree, with some strains causing syphilis, while others live a benign existence in pig intestines where they digest carbohydrates. The strains found in the Matses and Tunapuco are most closely related to the pig gut *Treponema*, which has led the study team to suggest they may play a similar carbohydrate-digesting role in the human hunter-gatherer gut as they do in pigs. What does that mean for the WEIRD people? Is it a problem to have a colon with about as much diversity as the average office of a Wall Street bank?* Some people definitely think so, including one of the authors of this study, Professor Christina Warinner, who has said gut microbiota diversity is 'critical to maintaining versatility and resiliency in the gut'. Professor Warinner has also suggested a lack of diversity might make us prone to inflammation in the gut, a key feature of intestinal problems like Crohn's disease.

Certainly some people have pointed to the rising tide of allergies in the West and blamed it on a lack of gut microbe

---

* Incidentally, a 2014 study in the Proceedings of National Academy of Sciences found that ethnically homogenous groups of Wall Street traders were more likely to be linked with price bubbles than diverse groups – possibly because they were more likely to agree with each other and reinforce inaccurate perceptions of asset values. So there is evidence to support increased diversity in financial institutions, for anyone who needs reasons beyond things like equality.

diversity, which they attribute to facets of the WEIRD life such as the use of antibiotics, processed foods and a passion for purging anything unhygienic from the environment. This is a compelling and fairly plausible narrative, but we're still learning what causes variation or a lack of it. One surprising discovery on this front has come from Elise Morton and colleagues from the University of Minnesota, who studied the microbiota of people in rural Cameroon. They found that one of the biggest predictors of high gut microbe diversity was the presence of the parasitic *Entamoeba*. These tiny single-celled parasites usually cause no problems, but some types can cause amoebic dysentery, which leads to severe diarrhoea. In the study, they found they could look at the person's mix of gut microbes and predict with 79 per cent accuracy whether they had *Entamoeba* in their gut. Why *Entamoeba* was associated with this greater diversity isn't clear – perhaps its presence stimulated the immune system and induced inflammation, which can shift the balance of microbes. Or maybe the *Entamoeba* wasn't the cause of the diversity at all, it just happened to prefer to live in a multicultural colon. We don't know.

Before anyone rushes out to eat amoebae (my friend who contracted amoebic dysentery would categorically advise against that), or indeed sloth or alligator or guinea pigs,[*] it's important to note that, as yet, there is no conclusive evidence that maximising gut microbe diversity is a worthwhile endeavour. In truth we don't currently know what a healthy microbiome looks like. Instead what's emerging is a picture of a complex system which is in almost constant flux and varies massively between perfectly healthy people. This was well demonstrated in a large study from the University of Michigan which looked at the microbiota of 300 healthy volunteers, and found a high degree of variability, not only between different individuals, but also within the *same* individual over 12–18 months. In mapping the variation across 18 different body sites over time, they found the mouth

---

[*] +/- tomato hat.

had the least stable community, like the microbial equivalent of transient squatters, while the vagina was the quiet suburban cul-de-sac of the map, with a relatively fixed mix of residents.

The relationship between these residents and food might be a two-way street – our dinner may influence our microbiota, and our microbiota may dictate our ability to digest our dinner. We don't have the necessary enzymes to break down certain sugars, and this includes porphyran, a sugar commonly found in red algae. Yet researchers rummaging through an international smorgasbord of stool samples found something which flies in the face of this fact: they found evidence of an enzyme capable of breaking down porphyran in the poo of Japanese, but not American, study participants. The enzyme belonged to a bacterium called *Bacteroides plebeius* which can be found on the seaweed commonly enjoyed by the Japanese. They concluded that the bacteria was digesting the porphyran and in doing so unleashing the nutritional benefits of the seaweed for itself and for the people eating it. If they're correct then this study suggests that our intestinal microbe portfolio can expand the number of foods we can gain sustenance from, allowing us to adapt to our environment at a pace that vastly outstrips human evolution.

Not all of our bacteria's eating habits have such positive perks – indeed for some genetically unlucky people their bacterial digestion can lead to the socially excruciating fish odour syndrome. When bacteria in our gut break down foods rich in choline (an essential nutrient found in things like eggs, fish and beans) they emanate a waste product called trimethylamine. In most people this is simply broken down in the liver, however, people with fish odour syndrome lack the enzyme necessary to break down the trimethylamine. It therefore builds up in their sweat, urine, saliva and vaginal fluid, giving off a smell described as being like faeces, garbage or rotting fish. An abnormal abundance of gut bacteria can also cause this syndrome in people who have the normal enzyme but can't keep up with the sheer scale of production of trimethylamine by their many bacteria. Unfortunately there is no cure for the condition, so those affected have to avoid certain foods or take antibiotics to thin out their bacterial population. The latter option brings the risk of antibiotic resistance so the antibiotics must be used infrequently, and for many keeping the disorder under control is a life-long battle.

Manipulating the microbiome

While most of us don't have to consider the olfactory impact of our diet to this extent, there is a growing market for influencing the microbiome for health and well-being benefits. However, given that we don't know what a healthy microbiome looks like, it's unsurprising that we don't know what we should be eating to cultivate the ideal bacterial collection. Some companies imply that a few gulps of 'good bacteria' will transform us into a state of perky perfection desired by all desperate housewives, but the truth is science is uncovering more wriggly questions than straight answers. For instance, there are studies in mice that show that dietary changes can lead to significant changes in bacterial metabolism within hours, suggesting that the microbes are highly responsive to the diet of their biological mansion. Yet giant pandas appear to have rather less responsive house guests. These endangered bears have been eating bamboo for over two million years and their microbiome still has more in common with the intestinal ensemble of a carnivorous bear, like their ancestors, than the herbivores they've become.

The abundance of wriggly questions makes it difficult to develop drugs that will allow us to battle diseases via manipulation of the microbiome. Yet as the success of the poo transplant showcases, it is possible. There are also other promising poo-related prospects out there, including the twenty-first century's holy grail: effective weight loss. A 2013 study in the hallowed journal *Science*, and covered with gusto in the mainstream media, described how obesity could be transferred via a human-to-mouse poo transplant. Microbe-free mice given the gut microbiota from an obese person put on weight, while those on the exact same diet but given the microbes from a lean person did not. They also found that the mice given the 'obese' type microbes did not gain weight if they shared a cage with those given the 'lean' microbes. This was attributed to the delightful mouse habit of coprophagy (poo eating), which effectively allowed self-administered poo transplants between the mice. It seems the result of the exchange of microbes between the mice was that the 'lean' microbes won out over the 'obese' ones. Of course, these were mice not people and there are many biological and environmental differences between us and them.* Also, the study found that a diet rich in saturated fats and low in fruit and veg prevented the 'lean' microbes from usurping the 'obese' microbes in the guts of the cage-sharing mice. So diet definitely matters and poo transplants are not yet a panacea for obesity.

Not everything in the microbiome has to involve poo. In 2013 Groeger and colleagues conducted a small trial involving people with three different diseases: 22 with ulcerative colitis, 26 with psoriasis and 48 with chronic fatigue syndrome (CFS). Their theory was premised on the idea that all three diseases have an inflammatory element, something that is widely accepted for the first two conditions though the nature of CFS is still the subject of debate. They tested all 96 participants for an inflammatory marker in the blood called CRP (C reactive protein) and two pro-inflammatory cytokines called TNF-alpha and IL-6. At the outset, CRP, TNF-alpha and IL-6 levels were significantly higher in all three diseases when compared with 22 healthy participants. Everyone in the study was then fed either a bacterium called *B. infantis 35624* or a placebo for six to eight weeks and their blood was then tested again. In all three diseases, CRP levels fell in those who had consumed the bacterium compared with the placebo. Overall, when combining the different disorders, 70 per cent of all patients given the *B. infantis* had a drop in their CRP, IL-6 and TNF-alpha, compared with just 9 per cent in the placebo group. Of course this was a small

---

* Not least the whole coprophagy thing.

trial and it was just looking at blood markers, so it didn't describe any improvement in the patients' symptoms, which is ultimately what we care about. However, it hints at a possible new line of inquiry for three very different but very serious diseases. That is hopeful, and exciting.

Stepping even further from the solid ground of established science fact and tentatively into the uncertain future is the prospect of treating mental illness through manipulation of the microbiome. This is a serious area of research, with the US government dedicating over $1million in 2014 to a new programme intent on investigating the microbiome-brain connection. As with obesity, much of what is known so far comes from mice, with the added complexity that diagnosing obesity in a mouse is rather more straightforward than complex mental health conditions like depression, schizophrenia or autism. With that 'caution' sign firmly planted, let's consider some of the evidence we have to date.

In 2013 a group of scientists found that mice with behavioural traits bearing similarities to autism (lower sociability, higher anxiety) had much lower levels of a gut bacterium called *Bacteroides fragilis* than mice without such traits. When they fed the mice *B. fragilis* the symptoms of autism in the mice stopped. They also found that the affected mice had higher levels of a bacterial metabolite called 4EPS, and that when 4EPS was injected into the mice without symptoms, they too developed the behavioural issues. Similarly, another group of researchers looking at rats found that feeding their rodents a daily probiotic cocktail of two bacterial strains produced a significant reduction in the rats' 'anxiety-like' behaviour. More compelling still was the fact that this same cocktail given to human participants also produced a significant reduction in psychological distress.

One key hurdle for microbiome-brain research is explaining exactly how the bacteria in our gut could be influencing our mood. A promising theory is that the mix of gut bacteria affects the integrity of the gut lining, something that seemed to be borne out by the fact that the autistic-type mice with low levels of *B. fragilis* also tended to have a leaky gut wall. A leaky gut could allow the transfer of chemicals from the gut into the bloodstream, which could travel to the brain and affect mood. Certainly we know gut bacteria can make neurotransmitters including the established mood-altering chemical serotonin, so this is one plausible pathway. Another is that a major nerve in the body, called the vagus nerve, provides a direct connection between the gut and the brain, and that neurotransmitters secreted by gut bacteria could be uploading mood alterations via this connection. Until we understand the mechanism we still have a long way to go before we can use it to benefit people with mental health issues.

While it will be some time before we start treating the microbiota to treat the mind, from all that we've learned in this chapter it seems clear the bacteria we keep company with matters. And therefore so do the people we keep close to us. A single 10-second French kiss is thought to transfer 80 million bacteria. Multiply that across years of dating and it's easy to see how our intimate partners can influence our microbiome. It also reminds us that we need an immune system to keep us safe from the bacteria-laden nature of intimate acts. Now, our jaunt through the immune system has been good so far, but if it's OK with you I think we're ready to take things to the next level. It's time for us to dance the immunological tango and explore the role of the immune system in sex, and possibly even in love.

CHAPTER SIX

Immunological Tango:
Sex and Love

It's not easy being a sperm. Propelled into an acid mucus bath and forced to swim past an aggressive border patrol that makes the immigration staff at Heathrow seem jovial, all to get to an egg that might never arrive. Yet we must remember that the vagina is a gauntlet for a good reason – it's a passage into the body, vulnerable to attack by all manner of delights, from syphilis to chlamydia. This most delicate of parts is therefore exceedingly well guarded at all times, but it's during sex that things get really interesting.

While you're popping on the mood lighting and Barry White, the first defence served by the vagina is acidic vaginal mucus. A healthy vagina has a pH below 4.5, which puts it somewhere in the league of acid rain, creating a highly unfavourable environment for sperm and sexually transmitted infections alike; research suggests that the lactic acid in the mucus slows the movement of HIV through it by up to 100 times. The low pH is also capable of immobilising sperm, but they are shielded by the seminal fluid they are ejaculated with. This fluid has a pH above 7.2, akin to the white of a freshly laid egg. Within eight seconds of hitting the vagina it can neutralise the vaginal acid, and completely change the pH landscape to a balmy pH 7.

If, like me, your mind rushes at this point to the questions polite people shouldn't ask, you'll be wondering 'and how exactly do we know this?' Science is full of intrepid explorers, and none more so than sex researchers. I remember well my admiration in student times for one esteemed Cambridge professor, who told me about his humble beginnings as a junior researcher collecting data created by amorous and actively mating sheep. That, though, was tame compared to Laud Humphreys, a sociology student in the 1960s who chose male–male fellatio in a public restroom (otherwise known as 'tearoom sex') as the topic for his PhD dissertation at Washington University. Humphreys waited in restrooms and volunteered to be the 'watchqueen', which involved alerting participants if the police or a stranger were approaching. This allowed him to observe hundreds of acts, and having gained the trust of participants he was able to explain his research, and interview many of them. Of course, some were unwilling to be interviewed, so to avoid losing valuable data from these reticent men he followed some of them to their cars, noted their number plates and turned up on their doorsteps a year later claiming to be a health service interviewer. He then asked them about their marital status, jobs and other details, never revealing who he really was.

Humphreys' research broke new ground by rebuffing stereotypes held in the 1960s about men who engaged in 'tearoom sex'. He found that 54 per cent were married and living with their wives, and 38 per cent were neither homosexual nor bisexual. These men were seeking quick, impersonal orgasms that could not threaten their public role as husband and father. However, these results were not the main legacy of this project – it stands as a case study in ethically unforgivable research. Even at the time Humphreys' approach led to uproar at Washington University, with a fist fight between faculty members, and half of the staff leaving the department in outrage.

Happily our pH evidence is not ethically questionable, and comes from a study that gathered data from a couple in the midst of coitus. This necessitated insertion of a glass pH electrode, a transmitter circuit and a mercury battery into the vagina, and leaving it there for two hours. How normal can sex be with all of that in the mix? In fairness the equipment was only 2.5cm (1in) long, but still, glass and mercury aren't standard sex toy ingredients for a good reason;* kudos to the methods section of the study for the boldly optimistic phrase 'its use intravaginally is quite acceptable'. Perhaps the set-up wasn't quite as acceptable as the authors thought, as they found themselves 'hampered by the difficulty in obtaining experimental subjects'.

Thankfully there *are* people willing to participate in this type of study. Such fearless researchers and study participants have led to the development of contraceptive gels that immobilise sperm by keeping the vagina acidic after sex. However, these modern gels aren't the first contraceptives to maintain the vagina's low pH. Four thousand years ago, the Egyptians used a pessary of seed wool, acacia, dates and honey as a sticky contraceptive solution that acted as a physical barrier with pH-lowering properties. To a modern girl this might sound like the inappropriate placement of a luxury bath product, but it's a significant step up from the contraceptive advice of the ancient Petri papyrus: a pessary of

---

* Although mercury and sex aren't entirely unacquainted. Until the twentieth century mercury was the main treatment for syphilis, leading to the phrase 'A night with Venus, a lifetime with Mercury'. Alas, much like syphilis, exposure to mercury could lead to brain damage and death.

crocodile dung, ant paste and honey. That milkshake wouldn't bring *any* boys to the yard.

Returning to the present day, while the acidic vagina is a challenge for sperm, the penis spares them from having to navigate the entire vagina (2,200 times the length of a single sperm). Unfortunately, the penis is like a slightly underfunded public transport service, taking the sperm a fraction of the way to their ultimate destination and dropping them prematurely in an unsavoury neighbourhood. This is the cervix, gateway to the uterus, and a veritable hotbed of immunological activity. It's rich in white blood cells including macrophages and neutrophils, which spew out chemicals such as proteases, which break down the invading sperm so they can be devoured and destroyed. The sperm have numbers on their side, though – at least 39 million, to be exact.* They're all deposited at once, which makes it difficult for the white blood cells to take them all on before some pass through the cervical opening and on into the uterus. This is more than a numbers game, though; once again, the seminal fluid supports the sperm in their endeavour. Amazingly, scientists have uncovered evidence of a careful immunological tango between male and female during sex, which is choreographed by the 900 proteins found in seminal fluid. We're beginning to appreciate the roles of these proteins, including protecting sperm from attack and helping to create a favourable environment in the womb for the embryo to implant.

One of the chief sperm-protection molecules is thought to be prostaglandin E, an immune messenger found in semen. Human semen is one of the richest biological sources of prostaglandins known to science. However, understanding the mechanism of sperm protection is an emerging endeavour. One theory is that prostaglandins reduce the rate of sperm capture by white blood cells by stopping them from adhering to the sperm. While prostaglandin's role in conception is still

---

* Impressive, but not as impressive as in pigs. One of the giants of ejaculation, a boar releases up to 100 billion sperm in a single load of 150–500ml (5–17fl oz) of ejaculate.

being explored, we know that it plays a major part in the less-fun process of labour. Prostaglandin gel is used to induce labour in overdue pregnancies, and the high concentration of prostaglandins in semen suggests there might be a biological basis for the urban legend that sex is a good way to get labour going. The Cochrane Group, which brings together lots of studies to assess the overall evidence for a theory, examined this issue. The only study they found included just 28 women, and reported little data. Cochrane suggested much more data would be needed to assess the benefits of sex in overdue pregnancies, but they recognised this might be easier said than done, with the splendid line 'it may prove difficult to standardise sexual intercourse as an intervention'. Quite.

The goal of sperm isn't simply to fertilise the egg. Success demands that the resulting embryo is able to implant in the uterus and thrive. Recent research suggests that semen may play a part in ensuring this through another immune messenger found in seminal fluid, TGF-beta (transforming growth factor). TGF-beta is a master at immune system manipulation, with potent anti-inflammatory and immune-suppression skills. This is where the human data ends and the mice take over (there's a limit to how intrepid study participants are). In mice, the TGF-beta in semen is released in an inactive form that becomes activated when it arrives in the vagina. This activated form of TGF-beta triggers an increase in the numbers of a white blood cell called the regulatory T cell, or 'Treg' to its friends. Tregs are the prefects of the immune system, suppressing the activity of other immune cells to maintain order and avoid friendly fire damaging the body's own cells. The Tregs recruited to the uterus help ensure that the embryo, which after all is a 'foreign invader', is able to safely implant. Treg deficiency in mice can lead to embryos failing to implant, and in people it is linked to unexplained infertility and miscarriage. This, combined with the high level of TGF-beta in human semen, and evidence that TGF-beta triggers an immune reaction in the human cervix, has led scientists to propose that TGF-beta and Treg might function in people in the same way they do in mice. If this could be demonstrated conclusively it would be a very valuable piece in the immunological sex puzzle, opening up possible avenues of investigation for infertility treatments and miscarriage prevention.

Awkward allergies

Unfortunately, the sperm survival kit can sometimes throw a spanner in the works: for example, some women are allergic to seminal fluid. While this is a rare allergy, a significant number of women with this affliction discover it the very first time they have sex. First-time sex is often more fumble than fireworks, but semen allergy makes a fumble sound positively magical; a swollen, irritated, burning vagina is not what every girl dreams of. For some women the allergy can even be life-threatening, causing breathing difficulties and loss of consciousness. Also, pity the poor boy whose semen is the offending cause – no lads' mag can prepare you for that eventuality.

There is uncertainty about which of the 900 proteins in semen is responsible for the allergy, though there is probably more than one culprit in the seminal chemical soup. Semen avoidance (i.e. using a condom) can completely prevent symptoms. However, this isn't an acceptable lifestyle choice for many women, particularly those who want to conceive. Semen allergy doesn't have a direct effect on fertility but couples do have problems conceiving because they are understandably wary of having sex without a condom. Unfortunately for these couples, the allergy isn't that well understood, and there's no clear consensus on how to treat it. Antihistamines 30 minutes before sex, anti-inflammatories and even semen protein injections have all been tried, with varying degrees of success.

However, at least women can avoid semen. In 2011 a flurry of headlines described a mysterious new syndrome of men suffering flu-like symptoms after orgasm. The source of these stories was a paper published by Dutch researchers which detailed 45 cases of Post Orgasmic Illness Syndrome (POIS). The men in the study ranged in age from their 20s to their 60s and many had been living with the symptoms of POIS for decades. Imagine having headaches, fatigue and fever an hour after every orgasm, knowing it will take two days until the symptoms resolve. It's no wonder three of the men had decided to abstain altogether, and eight could count the time between sex in months, despite having a long-term partner. Avoidance was harder among younger men, who found abstaining from masturbation and spontaneous emissions to be 'practically impossible'.

The cause of their symptoms is the subject of speculation, but the Dutch researchers suggest the most likely explanation is that the men are allergic to their own seminal fluid. They argue that the immune system fails to recognise semen as a self-made substance and whenever semen rushes along the inner tubes of the penis this triggers production of the antibody IgE (Immunoglobulin E), which recruits aggressive immune cells like T cells. This is one possibility, but ultimately POIS remains something of a medical mystery, with more questions than answers in the literature. For instance, if elevated IgE is responsible, how do we explain the absence of consistently elevated IgE among the patients? Frustratingly for sufferers, some of the biggest questions lie around treatment. The Dutch group reported that two patients had significant improvements following a 'hyposensitisation process' of having their own semen injected under their skin over several years. However there isn't a standardised treatment that has proved successful in a large number of patients. Much more research will be required before we understand POIS, which has received limited funding to date. In 2013 one patient told his story on Reddit, a social news website, and helped raise money for a modest research grant that may shed a little more light on the mystery. Until then, with no established treatment, patients are left with few options, with some resorting to drastic measures such as castration and homeopathy.

Sexed to death

Men with POIS undoubtedly have a life-altering affliction, but thanks to immunology, sex for the brown antechinus,

Antechinus stewartii, is life-ending. Within two short weeks this bug-eyed mouse-like marsupial goes from being a virgin to mating so much that its hair falls out, it gets covered in scabs and dies. This bizarre mating strategy is known as suicidal reproduction and is exhibited by four different genera of Australian marsupials, *Antechinus, Phascogale, Dasykaluta* and *Parantechinus*. The males in the various species included in these groups all have just one mating season in which to make some babies and secure their lineage. At the tender age of 10 months the brown antechinus stops making sperm and his testes begin to disintegrate, leaving him with storage tubules filled with sperm in need of a new home. So he starts mating with every female he can get his paws on, putting even the most dedicated porn star to shame with frenzied 14-hour sex sessions.

This is an understandably stressful time for the boys, and they barely stop their all-encompassing orgies to eat or sleep. Their bodies are consequently flooded with stress hormones, including cortisol. The chemical structure of cortisol is based on cholesterol, and it's this fat-rich make-up that allows this molecule to slip easily through the lipid membrane that surrounds every cell in the body, making it a powerful and far-reaching messenger of stress. Once inside a cell, cortisol is able to switch genes on and off. In the case of the immune system, cortisol switches on the production of anti-inflammatory chemicals and represses the production of pro-inflammatory chemicals. Intense suppression of pro-inflammatory molecules like IL-12 means that when the male antechinus is injured or gets an infection, his white blood cells don't get the message to turn up and fight it; in other words, the male starts to fall apart. Yet even when they are thoroughly ulcerous and gangrenous the males still 'run around frantically searching for last mating opportunities' according to antechinus researcher Diana Fisher, though by that point the females start to avoid them. You would, wouldn't you? The poor little male, sexed to death, eventually succumbs to injury and infection; they all do. The entire generation of males dies en masse.

Why? Well, the long-held explanation for this type of behaviour is that there isn't enough food available to sustain both the new and the old generation, so the males sacrifice themselves to ensure food for their offspring. However, Fisher believes this is an outdated and inaccurate explanation. She argues that natural selection acts at the individual level, whereas this altruistic explanation suggests the entire population is acting in concert. Her theory is that this behaviour has evolved because it gives a male's sperm better odds of success when competing against other sperm. The males of some animals compete to father offspring using teeth or claws or antlers, but in the Brown Antechinus it's down to the sperm to fight it out. And the more sperm you put out there, the better your odds of being one of nature's winners and passing on your genes. So nature selected for those most committed to the cause of mating, and nothing focuses the mind quite like your testes disintegrating and your subsequently having only one shot at passing on your genes. And lo, suicidal reproduction came upon the world. Let's just be grateful humans weren't blessed with that strategy.[*]

Slugs and snails versus sugar and spice

The stress hormones that so affect the little antechinus aren't the only ones with knock-on effects for the immune system. Sex hormones have been shown to have a significant impact on the immune system in people, leading to differences between men's and women's ability to fight disease. Viruses, bacteria, parasites and fungi: men are more likely to be infected by all of them, and when they catch an infection it tends to be more severe than in women. So, ladies, perhaps man-flu has a scientific basis after all.

One key reason for the male/female difference in immune response is that the sexes have different levels of hormones.

---

[*] Or the 'traumatic insemination' approach of bed bugs, where the male stabs his needle-like paramere (the bed bug equivalent of a penis) through the female's abdominal wall and sprays his semen inside her.

Testosterone, which is found in higher levels in males, tends to suppress the activity of the immune system. It reduces the production of antibodies and white blood cells (including T cells and B cells). Vitally, it also seems to reduce the expression of a receptor called TLR4 (toll-like receptor 4) on the surface of macrophages. The macrophage's TLR4 receptor is essential to its job as a first-responder to infection because it allows the macrophage to detect infectious agents, including bacterial toxins. By reducing the number of TLR4 receptors on the surface of macrophages, testosterone reduces the ability of the immune system to detect and respond to infection.

The effects of oestrogen, which is primarily a female sex hormone, are mixed. Oestrogen increases the production of antibodies, but it also reduces the production of some pro-inflammatory chemical messengers. One thing that does advantage women immunologically is that they have two X chromosomes, as opposed to the X and Y that men have. The X chromosome carries about 1,100 genes, some of which are responsible for producing vital building blocks of the immune system. This includes production of the receptors that cells need to be able to respond to an immune messenger chemical called Interleukin 2 (IL-2). A mutation in the IL-2 receptor can lead to a rare but serious genetic disorder called X-linked Severe Combined Immunodeficiency Syndrome (XSCID). We'll hear about this disease in depth in Chapter 13, but essentially children born with XSCID are very vulnerable to infection because they only have about 5 per cent of the T cells they need. This means they often begin catching infections in the first few months of life, and without a successful bone marrow transplant this syndrome is almost always fatal before the age of two. What separates the girls from the boys is that having two X chromosomes means girls have two copies of the IL-2 receptor gene. If one copy is mutated, they still have another functioning copy, so they won't have the symptoms of the disease. Normally female cells 'deactivate' one of their X chromosomes because they don't require both. Which X is inactivated is usually random, so that across all the cells in the body there is a roughly even split of active Xs. However, T cells in females with IL-2 receptor mutations seem to disproportionately deactivate the X with the mutated gene. This type of selective X deactivation also protects females from several other genetic disorders.

It's not all good news for girls, though. Women are more susceptible to sexually transmitted infections like HIV. Some of the key reasons for high infection rates among women are social. For instance, older men having sex with multiple young women is thought to explain the higher HIV rates among young women compared with young men in some cultures. However, there is also a biological basis for differences in transmission, with women twice as likely to catch HIV from an act of vaginal–penile intercourse as their male partner.

Women are also more likely than men to develop autoimmune diseases, where the immune system mistakenly attacks the body's own cells. About 5 per cent of people worldwide are affected by an autoimmune disease, and 78 per cent of these people are women. Systemic lupus erythematous

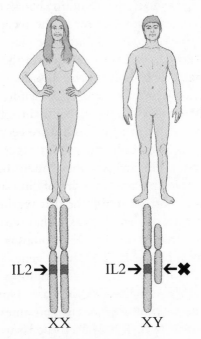

Figure 6.1 Inheritance of IL-2 receptor gene.

(SLE) is a classic example of this phenomenon, with women representing 90 per cent of all lupus cases. Lupus derives its name from the Latin for 'wolf', because thirteenth-century physician Rogerius thought the facial sores were reminiscent of a wolf bite. The effects of lupus can reach into almost every body system, causing everything from a rash to fever to seizures and psychosis. These symptoms tend to worsen during pregnancy and improve after the menopause, hinting that sex hormones play an important part in the autoimmune response. This theory has been supported by reports of lupus flare-ups being triggered by use of synthetic hormones, and the fact that lupus rates in men are similar to pre-pubertal and post-menopausal women. Moreover, men with Klinefelter's syndrome (where an extra sex chromosome is present, in this case XXY) are more likely to develop lupus than men with the standard XY sex-chromosome set-up.

However, the fundamental cause of lupus remains a medical mystery, so much so that it gathered iconic status in the TV series *House* as every junior doctor's best diagnostic guess when faced with a perplexing case. It was never lupus. Except once. One of the most popular theories for why the body turns on itself is that the disease begins with cells dying more frequently, and that in the process of death they display on their surface proteins that are normally hidden in the nucleus of the cell. Abnormally functioning T and B cells then begin responding to these proteins as targets to be attacked. Certainly, one of the hallmarks of lupus is the presence of antibodies against DNA. However, these antibodies often appear years before the person has any symptoms of lupus. So there are a lot more pieces of this puzzle we have yet to uncover. Where's House when you need him?

Monkey glands for millionaires and the modern penis

Time for a change of scene. It's the early 1920s; the world overflows with sumptuous jazz, flapper girls and Art Deco cool. You're a rich older gent with money to burn and an eye for the ladies – what's the one thing you need? A slice of monkey testicle implanted in your scrotum. Seriously. 'Monkey gland transplants' were all the rage in the early 1920s. They were the brainchild of Russian surgeon Serge Voronoff, who expounded the view that monkey hormones could return the vigour of youth to ageing men. A magazine article from 1923 describes how Voronoff would lay his conscious patient and an anaesthetised monkey on adjacent beds in the operating room. One of the monkey's testicles would then be carefully excised and sliced with the precision of a Michelin-starred chef into six delicate slices. These slices would then be popped through an incision in the man's numbed scrotum and nestled next to his testicles. The slices had to be thin to allow the man's testicular cells to 'interpenetrate' those of the monkey. Once man and monkey testicle became one, the hormones could begin their 'beneficial flow'. And what a flow. The benefits of monkey gland transplantation were every shopping channel's dream: incredible hair regrowth! improved eyesight! and muscles with 'spring'!

Voronoff's work was seen as a legitimate miracle by the establishment. At the height of his fame he was applauded by 700 of the world's leading surgeons at an international congress in London, and his patients were rumoured to include statesmen, actors and millionaires. Other surgeons began spin-offs of the operation, tinkering with testicles in all manner of weird ways and making ever more fantastical claims to boot. Dr Peter Schmidt, for instance, performed something akin to a vasectomy (no monkeys required) and claimed that repeated operations might stave off death itself. The holy grail of immortality lay at the end of a vasectomy: hmmmm. These wild claims didn't appear to ring alarm bells, and demand for Voronoff's original procedure was such that he reached out to the Governor General of French West Africa to arrange a monkey-breeding programme to secure a steady stream of testicles.

The testicles are an immune privileged site, which means they are largely a no-go zone for the immune cells that would typically reject foreign tissue. This fact has led scientists to investigate whether transplanting testicular cells can improve the survival of tissue transplants between species. In particular, there is significant interest in finding a way to transplant insulin-producing cells from the pancreas of a pig to a human diabetic to cure the person of their diabetes. Researchers in Mexico tried transplanting pig testicle cells at the same time as pig pancreas cells, all protected in collagen from the patient's own body, and concluded that the combination of collagen and the testicle cells had extended the survival of the pancreas cells. These pancreatic cells actually produced insulin inside the human patients, reducing their reliance on injecting insulin. So perhaps Voronoff's view that monkey hormones coursed through the veins of his patients might not have been so far-fetched after all ... However, even with all the technical capability of modern medicine, the Mexico study results were patchy at best, with only 6 out of 12 patients benefiting, and only limited evidence that the testicle cells suppressed the immune system. When it comes to Voronoff's experiments, in all likelihood the monkey tissue would have died from the lack of a blood supply. Equally, tissue damage from the procedure could have allowed white blood cells into the testicular area to attack and reject the monkey cells. To my mind, this enhances rather than detracts from the *real* miracle: all in all, about 2,000 men paid Voronoff very handsomely to implant monkey testicles into their scrotal sacks. Give the man an MBA.

The modern man doesn't pay to have monkey testicles intertwined with his own. No, he pays to have a sheet of processed skin from a dead person wrapped around the shaft of his penis to increase its girth. Historically, men could have their own fat injected into their penis, but this produced uneven results, and nobody wants to pay for bumpy bits. So phalloplasty progressed to sheets of skin from cadavers and leftovers from living donors who've had tummy tucks. The grafts are placed under the skin of the penis and wrapped around the shaft, leading to quoted girth gains of about 1–4cm (0.5–1.5in) in circumference. What makes these grafts immunologically interesting is that the tissue within them is processed to remove all the skin cells, leaving only a supportive matrix. By removing the cells, the process effectively removes the normal trigger for the patient's immune system to reject the graft. This makes it a popular material for plastic surgery, and not just in penises. There are over a million such grafts hiding in lips, noses and penises across America today.

I heart your HLA

Looks aren't everything, though. To complete the picture, all we need is love. From Kate and Will to Kimye, all of the big romances of our time may have been brought together by their immune systems. More specifically, they may have selected each other on the basis of a cluster of genes known as the Human Leukocyte Antigens (HLA). HLA is the human version of the major histocompatibility complex (MHC), which is a family of immune system genes found in a wide range of species, from mice to stickleback fish. HLA is important because it includes the code for the molecular markers that identify our cells as our own. Three of the HLA molecules, known as HLA-A, HLA-B and HLA-C, take peptides (sliced-up samples of proteins) from within the cell, and display them on the cell's surface to be reviewed by T cells. As we saw in the transplant chapter (Chapter 4), this biological barcode lets the T cell know that it's looking at a self-cell that should be defended, not attacked. However, HLA is relevant to more than just transplants because if the presented mix includes viral peptides the T cell recognises, it tells the infected cell to self-destruct. HLA molecules come in different shapes, and some are better able to present certain peptides than others. I like to think of it as an epic dinner service at Downton Abbey, with a plethora of plates, platters, forks and spoons for serving different foods. To be able to serve the full banquet correctly you need a lot of dinnerware. Likewise, the more variety you have in your HLA molecules, the better prepared you are to serve whatever peptides might be squirrelled away inside your cells. So in theory, if you choose a mate who has a different HLA serving set from you, your offspring will inherit a diverse set of HLA genes, enabling them to display a broad array of human and viral peptides.

Now, it's easy to spot if someone is tall, dark and handsome, but how can you spot someone's HLA? Maybe we can smell it? This intriguing idea comes from the 'sweaty T-shirt' studies of Swiss zoologist Claus Wedekind. In one of these studies he recruited four men and two women and asked them to avoid perfume and tobacco, and wear a cotton T-shirt on Sunday and Monday nights for five weeks. These sweat-covered T-shirts were swiftly collected on Tuesdays and offered up to the noses of 58 women and 63 men. These smellers had 20 minutes with the T-shirt to rate it for intensity and pleasantness. They found that people were more likely to rate a T-shirt as pleasant if the wearer had a different HLA combination from them. Wedekind thought this supported the idea that we sniff out partners who have HLA combinations we lack.

The idea that we select people with different HLA genes remains contentious, and the actual mechanism for a nose to detect HLA is unclear. However, that hasn't dampened the enthusiasm of genetic matchmaking sites like Gene Partner in Switzerland and Instant Chemistry in Canada. Just pop a cheek swab in the post and they'll test your DNA to establish your HLA profile. Then, using this, plus more traditional measures, they'll find your perfect match. The science of this is on somewhat shaky ground, but my inner nerd loves the idea of a day when lonely hearts ads read: 'Fun loving HLA-A*33:03:01 seeks HLA-A*24:02:0 with GSOH and a mansion'. They never have a mansion. Sigh. However, should you find your perfect partner you may decide that the next step is to bring together your HLAs by baby-making. The resulting adorable parasite, as we are about to see, is a wonder of immunology.

CHAPTER SEVEN

What a Cute Parasite: Pregnancy and the Immune System

I have a creepy, beautiful parasite squirming inside me. According to the internet it's currently about the size of a kumquat and has X-files-worthy webbed paddles where its hands and feet should be. That's not *exactly* how I phrased my pregnancy announcement. Pregnancy is many things – nauseating, exhausting, irritatingly cheese-deplete* – but it is also a feat of immunological magic. There is ongoing scientific debate as to how this impressive enchantment is achieved, but there is no disputing that over 40 weeks the body remodels

---

* The pregnancy-cheese minefield is because some, but not all, cheeses come with a higher risk of infections that can harm the growing baby.

itself to cater to the little parasite's every need. Start as you mean to go on!

Much of the language traditionally used in the scientific literature to describe the immune system in pregnancy makes it sound as if the expectant mum is fighting a losing battle with a particularly sneaky and belligerent adversary. While that's basically how I imagine life as a new parent, some eminent researchers from Yale have suggested it's not a fair description of the complex interplay between the two parties during pregnancy. Instead of mum's immune system being suppressed and beaten by baby, they propose a complex collaboration between the placenta (which is made of baby's cells) and the maternal immune system. The picture they paint is one of a majestic dance of immune cells and messengers, carefully choreographed to ensure mum and baby live and grow in (relative) harmony together.

However, during the first trimester of this dance it's as if baby is all left feet and mum's toes get firmly trodden on. It begins when the ball of cells that makes up the new embryo finishes bumbling along the fallopian tube and finds a spot in the uterus to burrow into. It tears a hole deep into the lining of the uterus, forcing its way down through layers of tissue, like a feral animal that leaves a trail of inflammation in its wake. Mum's response to this assault sees various immune cells swarming around the invading embryo, but far from halting the embryo's progress, this response is actually key to the embryo's ability to make a lasting home in the womb. In studies of mice and humans, the absence of Natural Killer (NK) cells appearing in the womb has been shown to result in the failure of the embryo to get deep enough into the lining to reach essential blood vessels and the pregnancy ends before it's really begun. Thus, far from being suppressed, the maternal immune system is alive and kicking and making sure the embryo has a fighting chance by assisting it to get settled well inside the lining. Unfortunately for mum, all of this inflammation and tearing and dying of cells contributes to the first trimester being a pro-inflammatory state, which occurs at the same time that the body is experiencing the hormone equivalent of a kick in the guts. The result is that the initial 12 weeks of pregnancy is typified by exhaustion, morning sickness and a general world-weariness that is a long way from a pregnancy glow.

That glow (I'm promised) arrives in the second trimester, which is thought to be a period of change from a pro-inflammatory to an anti-inflammatory state, as the lining of the uterus has healed and the placenta is well embedded. In this phase the baby has developed all of its organs, waving goodbye to its web-like paddle-hands and hello to functioning kidneys and the ability to hear mum's heart and voice. With the core structure in place, the baby spends much of the second trimester growing rapidly into a mini-me with the sustenance gained from mum across the placenta.

How mum passes on her immunity

The placenta is often referred to as a barrier, but that's a bit like saying a laptop is a calculator or a smartphone lets you call people. Yes, it's good at that particular function but it's also: So. Much. More. The magnificence of the placenta is that it's like the most efficient fast-food joint in the world combined with a communications platform that makes social media seem like a blind carrier pigeon, and a security system so sophisticated that James Bond would sell his own granny to the Russians just to get to play with it for five minutes. It is the sole route for food and nutrients crossing from mum to baby, it co-ordinates a cacophony of immunological chatter, and while its security settings stop certain things from crossing, there's a lot that does get transferred, including immunity to some diseases.

This latter role, in which the placenta acts as a conduit for the transfer of antibodies from mum to baby, is one is of the most important immunological achievements of the placenta. The antibodies collected while the baby is in the uterus provide her with protection against things her fragile little immune system has never encountered before, but may face in the outside world. The clutch of antibodies we inherit this way, known as natural passive immunisation, depends on the diseases our mum has encountered during her lifetime. So my little sproglet is likely to receive antibodies against some forms of the common cold, but not against dengue fever or anything else considered exotic for Scotland. In time she will have to build up her own portfolio of protective antibodies because the protective spell of passive immunisation is short-lived, lasting a few weeks to months after birth. Breastfeeding is a way of topping up the protection after birth as antibodies can also be found in breastmilk, especially in the rich yellow colostrum made in the first few days after birth. The benefits of this protective breastmilk cocktail can be lifesaving, a fact demonstrated in babies who are born into environments where cholera is rife. Those who receive antibodies against cholera in their mother's milk can still end up with cholera bacteria in their gut but do not develop the torrential, life-threatening diarrhoea that typically accompanies their presence.

Once upon a time women from a wealthy background could ensure their babies received these benefits of breast-feeding without needing to engage in the boob equivalent of hard labour. In seventeenth- and eighteenth-century Britain, back when formula milk didn't exist, a good living could be made breastfeeding other women's babies as a wet-nurse, with some women able to out-earn their labourer husbands. In a twenty-first-century twist on this centuries-old profes-sion, women today can advertise their breastmilk on the internet, using services like OnlyTheBreast.com to peddle their wares and ship frozen milk all over the country. Like the wet-nurses of old, their customers include women unable to breastfeed their own baby, but unlike ye olde times, today's consumers of white gold also include bodybuilders and breastmilk fetishists. The going rate varies and the sales pitch often includes the age of the baby mum is normally feeding, her diet (organic, vegan, dairy or gluten free) and occasionally images of the source of the product and whether the women are willing to let customers imbibe the milk directly. Some women claim they have made £15,000 ($20,000) a year by becoming their very own milkmaids. There are also milk banks, where women can donate breastmilk for free for babies whose mums can't provide it. This is particularly beneficial for the delicate guts of premature babies, and unlike milk bought online, the prospective donors are screened for various infectious diseases and the milk itself screened for bacterial contamination.

Humans aren't the only species who transfer antibodies to their offspring in breastmilk – many mammals do it too, including the cows whose breastmilk we splash over a crunchy bowl of cereal in the morning. In fact cows don't pass antibodies to their babies using the placenta at all and instead put their efforts into transferring high doses of antibodies in their milk. This fact has led some to explore the possibility of immunising people against human diseases using antibodies in cows' milk, a concept termed immune milk. Immune milk products are created by exposing the cows to selected killed bacteria or viruses to get their bodies to produce antibodies against the diseases caused by those micro-organisms. The cows then release some of these antibodies in their milk, as they would do to benefit their calves, but instead of benefiting a baby cow, the milk is harvested by people and prepared for human use. To date there have been experimental trials of this approach to treat diarrhoea-causing infections including *Rotavirus*, *Campylobacter* and *E. coli*, but the results of consuming these products have been very mixed – some studies found it reduced the diarrhoea, while others demonstrated no significant benefit. More recently scientists at Melbourne University have been creating immune milk rich in anti-HIV antibodies, which were able to make human white blood cells attack and destroy the HIV virus in lab conditions. Instead of consuming the immune milk, the team's long-term intention is to use the antibodies to make a cream to apply to the vagina that will protect women from catching HIV. The concept of immune milk is an exciting and developing one; however there is a long and obstacle-filled journey between producing antibodies in cow's milk and actually preventing or treating disease in people. It's safe to say we have some way to go before we're pouring epidemic-preventing goodness on our cornflakes.

Fortunately we can already hijack the human placenta to protect babies. For instance, whooping cough is a potentially fatal disease that babies can contract before they are strong enough to receive their first vaccinations. If mum is vaccinated against whooping cough at 28–32 weeks* of pregnancy then her immune system is able to craft anti-whooping cough antibodies and send them across the placenta. These then circulate in baby's bloodstream and provide temporary protection, which can be replaced with a longer-term solution when baby reaches two months old and can be vaccinated directly. This approach has been used routinely in the UK since October 2012 and to date there is very little evidence of risk to baby or mum, and plenty of evidence of benefit. A study of 20,000 mums looked carefully for evidence of any impact on stillbirth, premature birth, foetal distress, caesarean delivery, low birthweight or renal failure and found that vaccination did not increase any of these outcomes. What studies of the vaccination programme have found is that babies born to mums vaccinated at least one week prior to birth have a 91 per cent reduced risk of developing whooping cough in the first few weeks of life. This massive reduction has led the vaccination programme to be adopted in the United States, Australia, New Zealand and several other countries. In the UK at present, about 60 per cent of eligible mums take up the opportunity to be vaccinated, a number which will hopefully rise as experience with the vaccine grows and awareness of the risks and benefits becomes more widespread.

Transferring antibodies across the placenta sounds like a wonderful thing, and generally it is. However, for some women, myself included, there is a tiny but real danger that they could harm or even kill the baby, thanks to a disease known as haemolytic disease of the foetus and newborn. This disease can happen when there is a mismatch in the blood groups of the baby's parents, with mum having a negative

---

* Up to 38 weeks has benefit, but 28–32 weeks is the ideal window to maximise antibody production.

blood group and dad having a positive blood group. So in my case my blood group is A negative, while my partner is A positive. The positive and negative signs attached to our blood groups are short for 'Rhesus positive' and 'Rhesus negative', and refer to the presence or absence of a marker on the surface of the red blood cells called the D antigen. In terms of matching blood for transfusions, the D antigen is the second most important marker on red blood cells after the ABO groupings. About 15 per cent of people are rhesus negative, which has essentially zero impact on your life unless you need a blood transfusion, donate blood or have an organ transplant, or you're pregnant. In the case of blood transfusions, being rhesus negative actually makes you a more desirable blood donor, especially if your blood group is O negative, which is known as the universal donor type because you can give red blood cells to anyone. This is because O negative people have cells that are like a blank canvas lacking the main markers that disrupt the immune system of anyone receiving the cells.

However, being rhesus negative and pregnant is not so desirable. If the developing baby is rhesus positive and some of its red blood cells get into the bloodstream of its rhesus negative mum, then mum's immune system swings into action and develops antibodies against the foreign antigen D. The escape of baby's blood cells into mum can happen because of an injury during pregnancy, or it can occur during labour, which by all accounts is a traumatic event for all concerned. If the only exposure is during labour it doesn't have an impact on baby number one, which has already made its exit, but next time mum is pregnant with a rhesus positive baby, her immune system remembers the antigen D and is primed to attack. This is just mum's immune system functioning as it should, but the result for baby can be deadly. The antibodies can cross the placenta and latch onto the baby's red blood cells, labelling them for destruction. The result is haemolytic disease of the foetus and newborn, where the baby's red cells are broken up, causing anaemia, jaundice and, in severe cases, brain damage or even death.

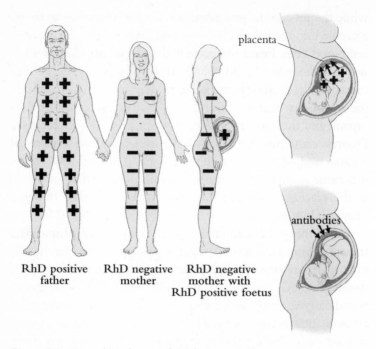

Figure 7.1 Haemolytic disease of the foetus and newborn.

Treatment of this serious condition used to involve drastic action: giving the baby a blood transplant while it was still inside the womb. In 1963 Sir William Liley was the first man to successfully pioneer this extreme solution. He had made three unsuccessful attempts on foetuses from different mothers before his fourth patient, Mrs E. McLeod, arrived at his office. Her first baby had been born healthy, but due to haemolytic disease of the foetus and newborn, her second died in the womb and her third pregnancy ended in the delivery of stillborn twins. Her fourth pregnancy would likely have ended tragically too but for the intervention of Liley, who successfully gave the baby a blood transfusion both in the uterus and shortly after birth, allowing it to live. What made this all the more remarkable was that he did it without the benefit of an ultrasound scan, which had yet to be invented. Instead early intrauterine blood transfusions relied on X-rays and dyes that would outline the baby's bowel to let doctors see where the baby was, and then needles were inserted through mum's abdomen to fix the baby in place while the precarious procedure took place. The aim was to get a needle into the baby's abdomen and flood it with the donated red blood cells. From there the red cells would wriggle their way into the baby's blood vessels and correct the severe anaemia caused by the antibody attack. There were many risks, including the possibility that the needle could puncture one of baby's organs, and the early success rate was poor at just 39 per cent. Progress came first in 1975 with the use of ultrasound to make it easier and safer to visualise the baby, and then in 1981 when it became possible to deliver the transfusion into a blood vessel in the placenta, avoiding the need to stab a needle into the baby's abdomen.

Today things have improved even further and intrauterine transplants are a last resort thanks to a preventative vaccination to stop haemolytic disease of the foetus and newborn ever developing. All prospective mothers have their blood group tested early in pregnancy. If it is negative, this fact is plastered all over their medical notes and they are given a vaccination at least once during the pregnancy and after any potential exposures to baby's blood. The vaccination contains a dose of anti-D antibodies. These manufactured antibodies latch on to any of baby's red blood cells that have made it into mum and effectively mop them up before mum ever has a chance to spot them and mount her own immune response. Thus her body has no memory of ever seeing rhesus positive red cells and any rhesus positive babies can safely snuggle inside her uterus without the risk of being attacked.

Infection across the placenta

Antibodies aren't the only thing that can cross the placenta and disturb baby's happy home in the uterus. Some bacteria and viruses can also travel from mum's bloodstream into baby's, with potentially devastating consequences. One example is *Listeria monocytogenes*, a type of bacteria that can cause birth defects and miscarriage, and sadly seems to share my love of unpasteurised soft cheese. In fact it appears I share a culinary palate with a range of unsavoury characters. There's also *Salmonella*, which likes a soft-boiled egg, and *Toxoplasma gondii*, which shares my penchant for parma ham and rare steak. On the plus side *Toxoplasma* can also be found in cat poo, so I may not get to eat rare steak but I do get a free pass on litter-tray duties for nine months.

In addition to causing miscarriage in pregnant women, *Toxoplasma* has the less distressing and more intriguing ability to relieve mice of their inherent fear of cats – like a super-unhelpful hypnotist with a vendetta against mice. While most sensible, law-abiding mice have an innate fear of cat urine, those infected with the single-celled parasite *Toxoplasma* are actually slightly attracted to the ammonia-laden eau-de-moggie. Even long after the infection has cleared and no parasite is detectable, the mouse's love of cat urine endures, a fact which implies that the parasite causes an irreversible change in the very structure of the mouse brain. There is method in the madness of *Toxoplasma*, which is thought to encourage this rampantly suicidal behaviour in an attempt to get its mouse host gobbled up by a cat. By using this mouse-shaped Trojan horse the parasite gets itself delivered directly

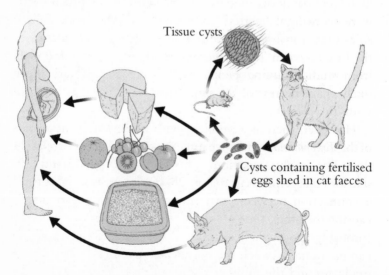

Tissue cysts

Cysts containing fertilised eggs shed in cat faeces

Figure 7.2 Toxoplasma's lifecycle.

into the cat's gut, which is where *Toxoplasma* likes to get it on for the sexual reproduction stage of its lifecycle.

In this context, meddling with the behaviour of the mouse makes sense, but it leaves us with the question: what does *Toxoplasma* do to the behaviour of humans infected by it? Most people don't develop a new-found love of sniffing litter trays, in fact for most people it has no impact at all. About 90 per cent have no symptoms, while 10 per cent develop vague complaints similar to a mild case of the flu. However, other stranger goings-on have been hinted at by some studies – for instance, one found an association between being infected with *Toxoplasma* and having a higher risk of traffic accidents. Another study of the blood samples of almost 46,000 Danish women who gave birth between 1992 and 1995 found those with *Toxoplasma* antibodies (which show they'd been infected at some point in the past) were almost twice as likely to have made a violent suicide attempt as those who had never been infected. Much smaller studies conducted at the Charles University in Prague focused on the psychological impact of *Toxoplasma*, and reported that infected men were more prone to being suspicious, dogmatic and jealous than non-infected men. The outlook was sunnier for infected women, who were more light-hearted, easy-going and willing to follow rules than non-infected women. More recently, evidence of the link between *Toxoplasma* and the human mind has come from studies finding people with schizophrenia are more likely than the rest of the population to have been infected with *Toxoplasma*.

However, there is a problem with interpreting the results of the *Toxoplasma* studies in people described above: correlation does not imply causation. This is a common mantra in statistics that reminds us that just because a statistical test finds a connection between two things, it doesn't mean one is causing the other. One of my lecturers at Cambridge used the (probably apocryphal) example that a statistical relationship can be found between the number of Cambridge University graduates and the number of prostitutes in Sydney. Even if true, this wouldn't be proof that Cambridge graduates are taking up job offers from Sydney brothels like they were management consultancy firms. There could be lots of other confounding factors that explain the apparent connection – for instance, perhaps there is a trend over time for an increase in both the number of students admitted by Cambridge *and* the number of prostitutes in Sydney because the population in both areas is increasing and demand is high. This doesn't mean one causes the other, it's just coincidence. In the case of *Toxoplasma*, some researchers have questioned the relationship with schizophrenia, noting that the prevalence of *Toxoplasma* infection varies significantly across the globe, while schizophrenia prevalence is fairly steady across different countries at 1 per cent. Instead it could be the other way around – people with schizophrenia are more susceptible than the general population to catching *Toxoplasma*. This is the idea of reverse causation, so instead of A causing B, it's actually B that causes A. For example, if a study came out showing pregnant women have a higher pickle consumption than the average person, we wouldn't conclude that pickles cause pregnancy. Instead we'd deduce that pregnancy causes pickle cravings. It's not always as easy as that to figure out which way round two things are connected, but we should always consider which is the most plausible order and look for supporting evidence in either direction.

To help pregnant women avoid food poisoning or potentially mind-controlling parasites, there is a wealth of guidance on what women should and shouldn't eat. Some of this is sensible, like avoiding mould-ripened soft cheese that hasn't been cooked. Even if the chances of catching listeria are incredibly low, that bit of cheese probably isn't worth the real risk of serious birth defects. However some of the advice is, in technical terms, bonkers. My partner calls it 'The War On Pregnancy' – the litany of instructions that predominantly serve to wear down women with a list of miserable things every good prospective mum should do or pay for, without providing an evidence base or sometimes even a basis in rational fact. These include advice on boosting your immune system with flu-fighting foods, which apparently include water. So, 1) things have reached a miserable low if water is a foodstuff, and 2) while keeping hydrated is good general advice, the notion that water will fight the influenza virus is worse than homeopathy.* Other gems include the need to drink protein shakes and instructions to swallow a pharmacy superstore of vitamins, minerals and what we shall term 'quacklements'. One example is Quercetin, which some websites recommend pregnant women should take to reduce allergy symptoms and food sensitivities. Other interesting claims for Quercetin include the protection of DNA, proteins and lipids from oxidative damage, and benefits to the cardiovascular system, liver, kidneys, and even mental state and performance. I can't find any evidence to support the allergy claims and a detailed investigation by the European Food Safety Authority in 2011 found no evidence to support Quercetin's other aforementioned 'talents'.

Waste not, want not

The wonders of the placenta are many, but I wouldn't fancy it with chips. Yet some women are very much OK with ingesting their placenta; Mayim Bialik (aka *The Big Bang Theory*'s Amy Farrah Fowler) suggests pureeing the placenta into a vegetable soup or blending it with fruits. She's not alone – there are a wealth of recipes on the internet, including roast placenta, placenta spaghetti bolognese and chocolate placenta truffles.[†] In 1998, British celebrity chef Hugh Fearnley-Whittingstall received a stern telling-off from the Broadcasting Standards Commission for serving a woman's placenta as pâté. He fried Rosie Clear's placenta with shallots and garlic, flambéed it, puréed it and turned it into a pâté which was served with focaccia bread to 20 of Rosie's nearest and dearest on live TV.

---

* Which at least tries to imply an active ingredient, even if it is water with a memory ...

[†] Though I'm not sure what you do if you don't have the required weight of placenta, it's not like you can just buy an extra packet. Not even Whole Foods stocks that.

She must have an incredibly open-minded group of family and friends; I'm not sure I could name 20 people who would eat my post-natal entrails. I'm almost certain hubby wouldn't be among them, and he definitely wouldn't show the gusto of Rosie Clear's partner, who tucked in to 17 helpings of the afterbirth pâté. However, not everyone watching was impressed, which is why the programme was chastised for breaching a taboo which 'would have been disagreeable to many'.

Maybe today it wouldn't cause such a stir because many a glamorous woman is indulging in a soupçon of placenta. *Mad Men* star January Jones had hers dehydrated and turned into pills, as did Alicia Silverstone – though in her case someone presented that to her as a gift.* Alicia has also suggested people can have a little piece of their placenta made into a healing tincture, which would certainly make an interesting addition to the first-aid box. This trend for placenta munching is predicated on the belief that it will impart a range of benefits such as increased energy, better milk production, better sleep, boosting of the immune system, warding off post-natal depression and even anti-ageing properties. However, the placenta is not sterile and some studies have actually found the placenta can harbour bacteria and elements like mercury and lead, perhaps related to its role in preventing harmful substances reaching the foetus. A 2015 review of the scientific literature on placentophagy (ingesting placenta) included 10 articles that studied humans or animals eating human placentas and found no conclusive scientific evidence of benefit and a lack of good data on the risks. They did find anecdotal reports from women who felt they personally had benefited from the experience, like the stories of Alicia and January, but such reports are very subjective. Any pregnant ladies still keen despite the lack of evidence and abundance of ick-factor can find illustrated step-by-step guides to placenta

---

* I find it difficult to keep Christmas presents a surprise, so I would love to know how they surreptitiously acquired her placenta to then give it back to her.

preparation and dehydration on the internet. However before stocking up on tincture of afterbirth or placenta pills it's worth noting Alicia Silverstone found the first 14 hours of labour 'almost sexy', so she perhaps enjoys a rather different perspective on life to most of us.

After nine long months of juggling the complex immunology of pregnancy, both mum and the placenta have done their job and it's time for a bit more inflammation: labour. Mum's immune cells flood the muscle wall of the uterus, creating an inflammatory environment that encourages the uterus to contract and deliver the baby, followed by the placenta. With a button nose and tiny toes, it's the most beautiful parasite that ever exits the human body. In our next chapter we'll meet the ugly ones, the sharp-toothed ones, the brain-eating ones. Let's reconvene at a palace for parasites.

CHAPTER EIGHT

A Palace for Parasites: Worms and Fleas and Ticks! Oh my!

Things lurk in the murky waters of the Amazon River. Swimming snakes, hungry piranhas and, according to some anecdotes, a tiny, spiny, blood-sucking fish with a fetish for the human male urethra. The candiru fish is the stuff of legend. The story goes like this: rather like sharks, which can detect tiny drops of blood in the ocean, the candiru sniffs out urine in the water and propels its slender body towards its beloved eau d'ammonia. The unsuspecting urinator has no idea what's speeding their way until the little fish launches itself into the stream of urine and swims up the golden arch and into the narrow opening of the urethra. Once inside it flexes out its spines and, like putting up a large umbrella in a narrow corridor, this lodges it in the shaft of the penis where it can do whatever small fish like to do when installed in a human penis. But therein lies the problem: a fish needs a human penis about as much as it needs a bicycle.

So why would the candiru bother? Perhaps the handful of stories on the subject were all a case of mistaken identity. Certainly the catfish family isn't renowned for its eyesight, so maybe the candiru was seeking some other animal's urethra and never intended to boldly go where no fish had gone before. But there are other problems with the apocryphal anecdotes, including laboratory evidence that the candiru doesn't respond to human urine and a notable absence of reports of candiru being found in large river animals like dolphins, which frequently urinate in the water. There's also the rather improbable physics of a fish swimming up a narrow stream of urine. Thus, on balance, the idea of a penis-loving parasite fish remains the stuff of fiction rather than fact.* But there are plenty of other nightmare-worthy parasites out there which are uncomfortably real, so before you sleep *too* soundly, let's venture into the palace of parasites.

A plague of little dragons

Dracunculus medinensis is one of the turduckens† of the parasite world – it's a parasite inside a water flea inside a person. This particular parasite causes guinea worm disease, once a scourge that affected millions of people worldwide but now confined to a handful of countries and a few hundred cases a year. The worm starts its story as a little larva hanging around in

---

* Unlike *Cymothoa exigua*, a very real crustacean that crawls into the mouth, eats your tongue and then uses its sharp little legs to wedge itself in place to become a beady-eyed alternative to the tongue it devoured. Thankfully the crustacean in question only does this to fish. So we're safe. Traumatised by the knowledge such a creature exists, but safe.
† For those unfamiliar with the splendour of a turducken, it's a dish made up of a deboned chicken stuffed inside a deboned duck stuffed inside a deboned turkey.

standing pools of water, where it looks like an easy meal for a peckish water flea.* Sadly for the water flea the parasite larva has more in common with a time bomb than a tasty snack ever should, and the treacherous morsel spends the next 14 days inside the flea preparing for the next phase of its life cycle: us. Eventually some unlucky person swallows water contaminated with the water fleas to create a temporary parasite/water flea/person turducken. However, this cosy arrangement is promptly ended by the plunge pool of acid that is the human stomach, which dissolves the flea. Unfortunately, instead of killing the parasite larvae, the acid simply unveils them, like unwrapping the world's most god-awful candy. The unleashed larvae then force their way through the wall of the intestines and into the spacious luxury of the abdominal cavity, before penetrating into the muscle wall of the chest and abdomen where they settle in and start to grow. After 100 days the larvae become mature worms and seek out a mate, turning the abdominal wall into their sexual playground. Having served their purpose the males then slink off and die somewhere in the body, while the pregnant females slide between the layers of muscle and begin to glide down towards their host's foot. It is a slow journey, during which time the female grows to a metre (over 3ft) in length with a uterus that takes up almost all of her body and which becomes heavy with 1–3 million tiny larvae.

About a year after the worm was first swallowed it looks like a giant piece of angel-hair pasta and it's ready to show its

---

* The types of hungry water flea involved sound like a cast of X-men rejects, including metacyclops, thermocyclops and mesocyclops. Mesocyclops is of note because as well as eating guinea worm parasites, it also eats mosquito larvae. This fact has led researchers to use mesocyclops to eliminate mosquitoes as a way of stopping the spread of dengue fever – a potentially fatal viral disease carried by mosquitoes in tropical and sub-tropical regions. Filling ponds with mesocyclops in a country with guinea worm parasites in the water would be a public health disaster, so this approach to dengue fever is limited to countries where the *Dracunculus mediensis* isn't in residence.

less than cherubic face to the outside world once more. But first it needs a compliant host to act as its midwife and help deliver its horde of babies into a suitably stagnant birthing pool. Its method for achieving this is the reason behind guinea worm disease's alternative name, dracunculiasis, which is Latin for 'affliction with little dragons'. The worm releases chemicals which cause a blister near the surface of the foot, accompanied by an overwhelming burning pain which the host desperately tries to soothe by plunging their foot into water. This is precisely what the worm needs, and it takes its cue to push its face through the blister and vomit its offspring into the water. The newly spawned larvae then await hungry water fleas of their very own and the cycle starts once more.

Dracunculus medinensis has been choreographing this parasitic paso doble for thousands of years. According to the World Health Organisation there are references to guinea worm disease in the Old Testament. In addition the Manchester Egyptian Mummy Project found mummified worms inside one of their embalmed Ancient Egyptians. The

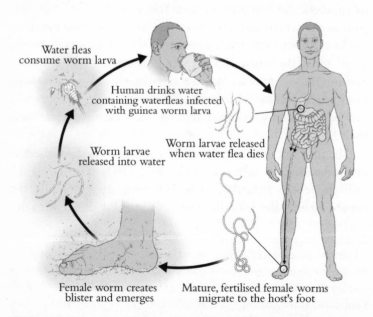

Water fleas
consume worm larva

Human drinks water
containing waterfleas infected
with guinea worm larva

Worm larvae released
when water flea dies

Worm larvae
released into water

Female worm creates
blister and emerges

Mature, fertilised female worms
migrate to the host's foot

Figure 8.1 Life cycle of Dracunculus mediensis (the cause of guinea worm disease).

longevity of our relationship with them is testament to the incredible capabilities of the worms, which have stayed ahead of the wave of new treatments that have ended so many other ancient diseases. We have no drugs or vaccines to effectively cure guinea worm disease. The mainstay of treatment is the same as that used by the ancient Greeks and Persians. When the worm pokes its head out you have to wind it round a stick and pull it out by a few inches every day until you've extracted all 100cm (40in) of white worminess. This old-school technique has led some to suggest guinea worm disease could have been the inspiration for the modern symbol for the medical profession, which is a snake, or arguably a worm, wrapped around a staff.*

Modern medicine isn't alone in failing to slay the little dragons – after all, the immune system allows the worm to live in the body for a year. This begs the question, how does the immune system fail to spot a metre-long worm growing inside us? Sadly the answer largely remains a mystery. One study found the worms secrete morphine, which is thought to suppress the immune system and stop the host noticing pain as the worm slides between muscles. However given all we know about the complex talents of the immune system, there must be much more than just morphine going on, because the worm does such a resoundingly good job that the body can't develop immunity. This means the same person can be repeatedly infected with guinea worm disease over their lifetime.

The immune-evading talents of guinea worm disease are so extensive that at its peak in the mid-1980s it infected 3.5 million people annually. However since then the tide has turned against the little dragons and in 2014 there were just

---

* Frankly either a snake or a worm would be a significant upgrade on the medieval symbol for doctors, which is thought to have been a urine flask. This was in recognition of the urine-observing, -sniffing (and sometimes -tasting) talents of the medical profession, or 'piss prophets' as one author describes them. Some even combined the analysis of urine with astrology in forming their diagnosis.

126 cases worldwide. This incredible reduction has been the result of a highly successful eradication campaign, which used techniques including water filters and insecticides to wipe out the crippling worm. Given the devastating impact guinea worm disease has had on countries,* this is a huge public health success. However it does mean there is limited time left to study the covert operations of this creepy little worm in its natural habitat. Understanding how parasites evade or subvert our defences isn't just important in helping develop drugs to kill them. It also helps us understand our own defence mechanisms, which in turn can help us create drugs to suppress or enhance the immune system. Yet we must not mourn the dragons, for there are many other parasites out there that have crafty ways and dark arts for us to study and fight.

The logic of brain worms

As any child knows, the key to a game of hide and seek is elementary: pick the best hiding place. In the human body, the best places to hide are those where the seekers (the immune system) find it hard to travel. This makes the brain a very smart place for a parasite to hide. Indeed plenty of creepy crawlies have been known to call the human brain home, lounging among our delicate little grey cells and causing some horrific symptoms in the process. This was the case for a 50-year-old man who had been troubled by headaches and noticing odd smells for four long years, causing his doctors to take regular scans of his brain. The results left the medical masterminds of Addenbrookes Hospital in Cambridge, UK, scratching their heads. Across the four years of scans they noticed bizarre ring-like patterns, looking much like Captain Scarlet's Mysterons, traversing 5cm (2in) across the right temporal lobe of his brain. So they did a brain biopsy and to the surprise of the surgeons they found the Mysteron was actually a live tapeworm by the name of *Spirometra erinaceieuropaei*.

---

* In Mali it's called 'the disease of the empty granary' as people are unable to work and harvest crops.

Since 1953 just 300 unfortunate people have had the honour of having this particular tenant ensconced in their grey matter. It doesn't actually eat the brain cells (it has no mouth so it couldn't even if it wanted to), instead it seems to suck up nutrients from its surroundings via its skin. It's normally found in crustaceans, amphibians and the guts of cats and dogs in China, so getting into a person was never part of the plan. This probably explains why it remained in larval form for the whole four years, which is just as well because the adult can grow to a brain-squishing 1.5m (almost 5ft). Nobody is sure how the worm infected this man, but it could have been through contaminated meat or water.[*] However the worm got in, it then wriggled up through his body to reach its cranial penthouse where it could enjoy the luxury of a very special hiding spot. As an immune privileged location the brain receives less attention from the scanning eyes of the immune system than other bits of the body. In part this is to actually protect the brain; the skull means the brain is in a confined space, so inflammation and the associated swelling of the tissue can cause more harm than good as the brain gets squashed against the sides of its bony box.

The brain's immune privileged status results in part from architectural peculiarities of its structure. For instance, a specialised mesh of cells coats the outside of blood vessels in the brain and acts like a form of border control, policing what's allowed to exit the blood and migrate into the brain, including limiting the access of white blood cells. The traditional burglar alarm of the body, the lymphatic system, is also less well hooked up in the brain. This superhighway for white blood cells involves an elaborate network of vessels, nodes and organs, which circulate a clear liquid called lymph whose main ingredients are white blood cells and a fluid called chyle that comes from the intestines. Fewer connections

---

[*] One quirky way of becoming infected with *Spirometra erinaceieuropaei* is when people use a traditional Chinese frog poultice, which involves applying raw frog meat to sore eyes to soothe them. I think I'll stick to cucumbers.

to the lymphatic vessels means a smaller number of tell-tale intruder clues are transported from the brain to key immune system hubs like the lymph nodes and spleen, reducing the probability that brain parasites are detected. In addition, the environment of the brain doesn't really create the kind of climate an essential white blood cell called a T cell can thrive in, because there's a shortage of its favourite things, such as immune chemical messengers that normally direct and encourage it. This isn't to say the brain is unguarded, the innate immune system is on active duty, but it does make the brain a more favourable hiding spot than the lungs or the liver.

The scourge of sleeping sickness

Some parasites take a belt and braces approach to hiding. Take *Trypanosoma brucei gambiense*, a single-celled parasite that looks a bit like a tiny worm that tapers from a chubby head and body to a whip-like tail. Spread by the bite of an infected tsetse fly, it causes 98 per cent of cases of sleeping sickness (also known as trypanosomiasis) across sub-Saharan Africa. Not content to hide in the brain, it adds another layer of complexity to its evasion strategy by changing its surface appearance faster than the immune system can keep up, a technique called antigenic variation.* It's as if the trypanosome has a bag of hats that it can whip out and use to play dressing-up to outwit the immune system. Initially all the trypanosomes in the body are wearing purple fedoras but by the time the immune system has figured out how to make anti-purple fedora antibodies, some of the parasites have had a rummage in their genetic dressing-up kit and popped on a pink bowler hat instead. The pink bowler brigade are immune to the anti-purple fedora antibodies and therefore persist in the body and multiply. The trypanosomes have about 1,000 hats at their

---

* Other disciples of antigenic variation include the malaria parasite and the bacteria responsible for the sexually transmitted disease gonorrhoea.

disposal, which means this cycle of deadly dressing-up goes on and on, allowing them to thrive for years.

However, each cycle of antibody creation and attack creates inflammation and damage to the surrounding cells. When the trypanosomes are in the first phase of infection they are in the blood, lymph and under the skin. Therefore this inflammation can go unnoticed for months or even years because it causes few major symptoms, except vague fever, joint pain and itching. However, once the trypanosomes cross into the brain the inflammation causes changes in behaviour, confusion and disturbance in the person's sleep cycle. Without appropriate treatment, sleeping sickness is almost universally fatal. According to the WHO, during the most recent epidemic over 50 per cent of the population in some villages in Angola, South Sudan and the Democratic Republic of Congo were infected. At the time this made sleeping sickness the first or second biggest cause of death in those communities, outstripping other devastating diseases like HIV.

Sadly, like guinea worm disease, the treatment options for sleeping sickness are limited at best. One of the main drugs for treating the first phase is suramin, a drug discovered in 1920 whose mechanism of action we still don't really understand, and that has a range of undesirable side-effects including cloudy urine and a crawling sensation in the skin. Still that's better than melarsoprol, one of the drugs for the brain phase of sleeping sickness, which was discovered in 1949 and is derived from arsenic. Its negative effects can include damage to the heart muscle and a generalised scaly eruption over most of the skin. Worse still, 5–10 per cent of people who take the drug suffer a brain reaction called reactive encephalopathy that is fatal in 50 per cent of cases. That means up to 1 in 20 people suffering from sleeping sickness treated with melarsoprol will be killed by the drug. That's a calculated risk no patient should have to take.

In search of a modern alternative to these old and toxic drugs, some researchers have turned their attention in a surprising direction: baboon's blood. Baboons, along with other primates like mandrills and the beautifully named sooty mangabey, are immune to all of the African trypanosomes. In their quest to explain this immunity researchers have found a protein in the blood called apolipoprotein L1 (ApoL1), which seems to play a key part in protecting the baboons. Moreover when this gene is put into mice, those mice become immune to a type of trypanosome called *T. b. rhodesiense* (which is responsible for the 2 per cent of human sleeping sickness not caused by the *T. b. gambiense* described above). In fact humans also have a version of ApoL1, which is about 60 per cent similar to the baboon version. Our version isn't normally effective against the two types of trypanosome that cause sleeping sickness; however some people have a rare mutation of ApoL1 that can kill *T. b. rhodesiense*. The fact that this mutation has only been found in people of African descent suggests it could have evolved because of this protective effect.

All of this has led to much interest in ApoL1 as a way of ending the scourge of sleeping sickness. But it's also led to a lot of heated debate as the top scientists in the field can't agree on exactly how ApoL1 works. They all agree it's a multi-stage process that involves other proteins that help smuggle the ApoL1 into the parasite. From there it travels to the trypanosome's equivalent of a digestive system and pokes holes in it, which leads to the demise of the parasite. However, there is still all to play for in providing definitive proof of exactly how ApoL1 does all this. The three leading scientists in the field all have very different and very strongly held views, fuelled in part by the fact that they can't seem to replicate each other's findings. Apparently their exchanges are the highlight of trypanosome conferences. Yet amid the drawing of pipettes at high noon, the important fact is that these brilliant minds are all working to understand and ultimately terminate trypanosomes – a reality the millions of people living with the threat of sleeping sickness dearly need.

Wuchereria bancrofti – a worm with a cloaking device

Some parasites have a partner in crime to help them win the eternal game of hide and seek they play with us. This is the chosen strategy of the roundworm *Wuchereria bancrofti*, which is a tiny thread-like worm that uses a bacterium to help cloak itself from the immune system. The worms are spread by mosquitoes that pick up microfilariae (immature worms) when they bite an infected person. The worms mature inside the mosquito until they are ready to return to a human host, at which point they are transferred from the mosquito to their victim's skin and make their way into the body. They travel to the lymphatic system, where they live for an average of six to eight years, growing into adult worms whose bodies are littered throughout the lymphatic fluid, resulting in the disease lymphatic filariasis. Sometimes there are no outward signs of the infection, but inside the roundworms are producing millions of baby worms, resulting in unseen damage to the kidneys and lymphatic system. If people do experience acute symptoms they tend to get bacterial skin infections that result from the damage to the normal immune surveillance role of the lymphatic system. In the long term, the infestation can stop the normal drainage of the lymphatic

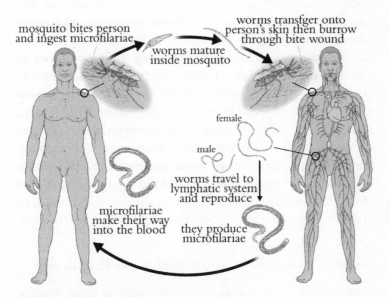

Figure 8.2 Life cycle of Wuchereria bancrofti.

fluid to the extent that the person's limbs can swell up and the skin on them thicken in a condition known as elephantiasis.

One other part of the body that can be hit by the inability to drain lymphatic fluid is the scrotum. This happened in dramatic fashion to a 20-year-old Haitian man whose infection with *Wuchereria bancrofti* caused his scrotum to swell over a five-year period. By the time he met the surgical team describing his case, his scrotal swelling stopped him walking or using his penis in the ways any normal young man might hope to. The swelling reached halfway down his leg, was the size of an angry sack of basketballs and weighed 55kg (121lb) – making it one of the largest cases of scrotal elephantiasis in recorded history. Following successful surgery to remove the huge mass, the wound healed over and the man was able to return to normality.

While a 55kg (121lb) scrotum may not sound like the work of a master of disguise, that is precisely what *Wuchereria bancrofti* is. And it's wildly successful at it, managing to live inside people for years without being cleared by the immune system. It currently affects over 120 million people in 73 countries across Asia, Africa, South America, the Caribbean and Western Pacific. Part of *Wuchereria*'s success is attributable to its partner in crime, a bacterium called *Wolbachia* that lives inside the cells of the worm. In fact *Wolbachia* keeps interesting company in general* – it enjoys residing in several different parasitic worms, including *Brugia malayi* (another worm responsible for lymphatic filariasis) and *Onchocerca volvulus*, which gets into the eyes and causes river blindness.

The relationship between the *Wolbachia* and these parasites is so long-standing and intimate that the worms now depend on *Wolbachia* to make a range of essential molecules, including

---

* *Wolbachia* also infect ladybirds which inherit the bacteria from their mother. The inheritance via the female lineage means *Wolbachia* has no use for male baby ladybirds, so it kills them before they've hatched. This gives the girls something to munch on when they hatch, improving their survival chances and therefore the chances that the *Wolbachia* will be passed on to the next generation.

the basic building blocks of DNA. The dependency seems to run both ways, as scientists have found *Wolbachia* is incapable of making a range of vitamins and molecules essential for its enzymes to function and relies on the worm to provide these.

Recent research suggests this little partnership perplexes the immune system, which isn't sure how to approach the bacteria/worm hybrid. Responding to the presence of molecules secreted by the *Wolbachia*, neutrophils surround the worm and begin breaking down the chemicals its bacterial side-kick has been producing. But these chemicals are like a glamorous magician's assistant shaking its tassels and flashing its feathers to distract attention away from what's really happening – in this case, the presence of a little worm. Not only do the neutrophils leave the worm alone, but their presence also stops other white blood cells called eosinophils, which are more adept at worm killing, from getting close.

In 2011 Benjamin Makepeace and colleagues conducted an experiment to break up the worm–*Wolbachia* union by treating cows infected with one of these parasite partnerships with some antibiotics. The drugs killed off the *Wolbachia*, and the neutrophils moved away and were replaced with eosinophils, which set about destroying the worms. In the group of cattle treated with drugs aimed at the worms themselves (rather than their bacteria) there was no change in eosinophil behaviour, which lends weight to the idea that *Wolbachia* is helping the worm to hide.

However, while the partnership may enhance the worm's ability to thrive, it's also a huge vulnerability because from a pharmaceutical point of view there are twice the number of targets available. You need only kill either the worm or its bacteria to end the disease. This has been exploited by scientists who have successfully trialled the antibiotic doxycycline to kill the bacteria in order to kill the worm. Unfortunately doxycycline can increase the skin's sensitivity to the sun, which isn't particularly practical in the hot countries most affected by these worms, so the quest for the optimal treatment for lymphatic filariasis continues.

Worm therapy

The examples we've encountered so far are just the tip of the parasite iceberg. There are flatworms, roundworms, hookworms, whipworms, fleas and ticks, lice and amoebae. They're all queuing up to get a room in the palace of parasites and the brightest brains on the planet are struggling to understand how they do it. Possibilities range from parasites that act as puppet masters for our white blood cells and direct our immune response down a losing strategy, to parasites with sartorial skills that craft themselves a human suit made from scavenged proteins, effectively rendering them invisible to the immune system. Yet all is not lost, because we have equally clever ways of killing them.

Chief among these is a class of antibody called IgE (Immunoglobulin E). IgE is normally produced in relatively small quantities, but comes to the fore during a parasite infestation. Some academics believe the types of response triggered by IgE, like itching, scratching, sneezing and coughing, all evolved to dislodge parasites that are too big to be eaten by cells like macrophages. As we'll explore in a later chapter, IgE also plays a part in allergic reactions by triggering the sneezing, itching response to harmless things like pollen. This has led to considerable interest in using parasites, with their impressive box of tricks for shushing the immune system, to treat allergies and other inflammatory diseases. To date two parasitic worms have been trialled in people: the pig whipworm and the human hookworm.*

---

* Sometimes you have to be pretty open-minded to be a patient – imagine how troubling your disease has to be before you say 'sure I'll swallow worm eggs' or 'yeah, maggots would be marvellous, thanks'. In 1824, one poor sailor was so distressed by his priapism (a prolonged erection without sexual desire, named after a Greek god with oversized genitals) that he let his doctor apply 20 leeches to his perineum. Between the leeches, having rhubarb applied to his penis and letting the surgeon stab his penis with a lancet, he went home cured, though unsurprisingly impotent for life. Worm eggs don't seem so bad by comparison now, do they?

The first forays into worm therapy began in the early 2000s and involved small clinical trials to assess the safety and efficacy of using pig whipworm in people with inflammatory bowel disease. Patients swallowed the parasites in their egg form, which travelled through the gut to hatch in the colon. The early results were promising as the treatment had no short-term or long-term adverse effects and resulted in a significant reduction in how active the patients' disease was. This spurred researchers on to do larger trials. Sadly, that's where things get a bit disappointing. In October 2013, the first larger trial with 250 patients announced it had found no impact on disease activity or remission rates. This was quickly followed by the early termination of another trial following the recommendation of an independent monitoring committee, which cited a lack of efficacy. Despite these disappointments, this is still an emerging field and with several hotly anticipated trials still ongoing, it's too early to call time on the pig whipworm debate.

It's also not the only player on the field. There have been trials using the human hookworm, although these have been rather more controversial. One of the wonders of pig whipworm is that the worms only live in the gut for a few weeks before the colony dies out. This does mean repeated doses are necessary, but also significantly reduces many public health qualms about purposely infecting people with worms. The human hookworm is a more problematic prospect, because once established the infection can last for years and the worms themselves cause symptoms like anaemia and digestive problems. It's also got quite an unpleasant commute to work as it gets applied to the skin, burrows into the blood vessels and travels to the heart and lungs. The immature worms then crawl up through the lungs to the back of the throat before being swallowed and transported to the small intestine, where they survive by sucking blood from the cells lining the gut. There are currently major global health efforts dedicated to purging the hookworm from people's digestive tracts; so plans to put them in there on purpose have received an understandably mixed response. This hasn't been improved by the fact that the results of the few small trials which have been conducted have also been rather mixed. One administered the worms to nine patients with inflammatory bowel disease and reported that seven experienced improved disease scores but two patients actually worsened. A slightly bigger study of 20 patients with coeliac disease (an allergy to gluten) found that, compared with placebo, those treated with the worms had lower levels of inflammatory chemicals in their gut. However, when they were exposed to wheat there was no difference in symptoms between the worm group and the placebo group, so it wasn't a resounding success.

Other studies have looked at the correlation between having a worm infection and protection against asthma. However a systematic review of 33 studies concluded that worms in general seemed to provide no clear benefit. The only hint of hope was that when you narrow your interest to hookworms specifically there was some evidence of protection against asthma, and that level of protection was greater with a higher density of worm infection. Yet, as with inflammatory bowel disease, actual clinical trials with hookworms have failed to demonstrate impact on the symptoms of asthma. This doesn't actually nix the idea that the worms can prevent asthma, perhaps just the idea they can *treat* it. If you look at studies of animals infected with worms there is a sizeable volume of literature to suggest worms can prevent the development of allergic reactivity, but only a handful to support the notion that worms can have an impact on established allergies. So perhaps we're just not taking the worms early enough? Maybe, but the animal studies also involve much higher doses of parasites than any half-decent human ethics committee would allow. So for now, worms as a medication remain a work in progress.*

---

* Unless you have access to the internet and a disregard for common sense. Then you can order the most glamorous of weight-loss strategies: a tapeworm tablet. While modern-day dieters have been known to order tapeworms online, the trend is thought to have been first advertised in the 1900s to help women squeeze into their finery. Not only is it conceptually unpleasant, it's also exceedingly dangerous as the worms can grow to over 9m (30ft) in length, which is large enough to physically block up the intestine.

In 2015, the Nobel prize for medicine went to three pioneers of anti-parasitic drugs. Eighty-four-year-old Tu Youyou discovered artemisinin, a drug that revolutionised malaria treatment at a time when the parasite was becoming resistant to the standard drug chloroquine. The other half of the prize was shared by William C. Campbell and Satoshi Omura for their work on a drug to combat roundworms. While these drugs were designed decades ago to destroy parasites, an exciting thought is that perhaps in future the Nobel prize will go to scientists who have made revolutionary drugs *from* parasites. As we build a deeper understanding of how parasites toy with us, we may be able to make synthetic equivalents to their box of tricks which can be used to soothe inflamed colons and ease asthmatic lungs. This is no real leap of imagination or flight of fancy, it's just progress.

The drugs of the future are based on the hardcore teasing apart of immunology of the past. One of the most important examples of this teasing apart has been developing our understanding of B cells and T cells. While the innate system protects us against invaders in general, it's our B and T cells that have a talent for bespoke assassination to follow this general response and put an end to any infection. From spiral-shaped syphilis to brick-like poxviruses and spherical particles of pollen, our B cells can produce tailored antibodies to bind them all. Our T cells, on the other hand, are the grand high inquisitors of the immune system, spotting and destroying infected cells and even cancer. We've met them in passing in some of the chapters so far, but in the next chapter we will become well acquainted with these assassins of the adaptive immune response. Crucially we'll learn how they are able to generate a targeted response to virtually any bacteria, virus or other potential immune system provocateur on the planet.

CHAPTER NINE

The Adaptive Assassins: Killing Everything from Pink-eye to Plague

Our bodies can make 1,000,000,000,000 different antibodies, which means that hidden inside each and every one of us is the capacity to make as many types of antibody as there are stars in our galaxy. This incredible breadth also raises a curious question: we only have about 30,000 genes to code a whole person, from nerves to fingernails, with a few hundred of those genes ear-marked for antibody production. So how can we make a trillion different antibodies from hundreds of genes?

How to build a galaxy of antibodies

Well, first let's consider what it takes to make just one antibody. An antibody is a large Y-shaped protein. The two arms of the Y are responsible for sticking to the target or 'antigen', whether that's a bit of a bacterium or a brain worm. On the other end is the tailpiece, which acts like a cattle prod to kick other immune cells and molecules into action to destroy the target. The Y shape is constructed from four protein chains woven together: two identical heavy chains and two identical light chains. The tailpiece is made from just heavy chains, but in the arms of the Y the light and heavy chains are intertwined, running up the length of each arm and combining at each tip. The tip is the bit responsible for making first contact with any intruders the antibody encounters and is called the antigen-binding site; it's therefore also the bit we need to vary between antibodies. We need to be able to create enough antigen-binding sites to be able to respond not only to every microbe-shaped threat that exists today, but also to every threat that might ever appear in the future. If pandemic pigmy hippo flu or sexually transmitted explosive diarrhoea were fact rather than fiction, you'd want an immune system that could cope, right? Then you need to make a lot of different antigen-binding sites, and the secret to that superpower lies in the many different types of heavy and light chains we can create. Each B cell makes its own unique antigen-binding site, and will only ever be capable of making that one shape its whole life. So let's make like a B cell and create a unique antibody.

First, we need a heavy chain. Antibody heavy chain proteins are made up of two distinct sections: a constant bit and a variable bit. As one might expect, the 'constant' bit codes for a part of the heavy chain that is pretty uniform across all antibodies. However, there are five different constant sections to choose from, creating five distinct classes of heavy chain which make five different classes of antibody, a subject we'll return to shortly.

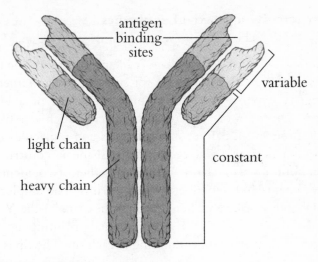

Figure 9.1 Antibody structure.

So we have five different antibodies, but let's face it, that's like having a small selection of pointy sticks and going off to take on a dragon, or actually a legion of dragons with big wings and fiery breath. We need more antibodies, which is where the variable region comes in. Unlike the constant region where a single chunk of DNA is translated to make a protein, the variable region consists of multiple chunks of DNA with some assembly required. The three necessary elements are: one variable (or V gene) segment, one joining (or J gene) segment and one diversity (or D gene) segment. We all have 51 different V segments, 27 D segments and 6 J segments, which can be joined in any combination so long as you have one of each in your variable region. One way of thinking about the heavy chain construction process is imagining you have to make a Mr Potato Head army, and you know that the more variety in your vegetable warriors the better. So you start with picking a head, your constant region, and now you need to give him a pair of eyes (V gene), a mouth (D gene) and a nose (J gene). You put your hand in a bucket with 51 unique sets of eyes and pull out a pair, then you reach into another bucket with 27 different mouths and grab a gob. Then finally you randomly select one of six distinctive noses. Combined with the five possible types of constant region this would give you 41,310[*] possible feature combinations for your fearsome Mr Potato Head battalion.

This is still a long way from the trillion antibodies we can make. However so far we've only made heavy chains, so next let's make a light chain to pair with it. This time there are only two options for the constant region, producing two different types of light chain called kappa light chains and lambda light chains. As well as having fewer constant regions to choose from, light chains also have more simplicity in their variable region, with only a V segment and a J segment required. So this Mr Potato Head has no mouth, and only 40 pairs of eyes and 5 noses to randomly choose from. You don't need a Fields medal to realise that combining every permutation of light chain with every permutation of heavy chain would *still* leave us short of a trillion combinations. But that's OK, because we're not done yet.

Normally cutting and pasting DNA segments is an eye-wateringly precise process. I'm talking about the kind of precision you'd expect from a person with joint degrees in bomb disposal and forensic accounting. However in the case of V(D)J joining, it's more like the level of exacting precision you'd expect of an amateur clown with butterfingers and a penchant for snorting laughing gas. The result is that throughout the process small bits of the ends of the gene segments can be lost or tiny clippings of DNA added. This random loss and gain during the pasting of the segments is called junctional diversification and it introduces a whole new level of variation to the creation of both light and heavy chains. It's rather as if the quality control people at the Mr Potato Head factory have gone out for the afternoon, and so nobody stops the green noses or broken toothy grins escaping off the conveyor belt of parts. Of course, this lax quality control has consequences – some of the V(D)J gene segments

---

[*] Just multiplying it up, so 27 x 51 x 6 x 5 = 41,310.

are too damaged to work any more. Indeed about 66 per cent of the gene segments produced this way are effectively junk. There's no point sending unfit Mr Potato Heads into battle, which is why the B cells that make non-functional antibodies never survive to exit the bone marrow.

The V(D)J joining is not an entirely lawless process, and there are a few key quality control points. For example the individual V, D and J gene segments are each flanked by DNA codes that help ensure that the right combination of segments are pasted together. Through this tag system, B cells can chop up their own DNA and still create a world where your V gene is connected to your D or J gene, not to another V. Another quality control point happens after the initial cut-and-paste job has been done. If the antibodies produced by the rearrangement stick to our own healthy cells they are prompted to undergo a rapid re-arranging of the V(D)J genes. If they fail to come to an arrangement that doesn't react to our self-cells* they are deemed too dangerous to live and are set to self-destruct. By employing this carefully cultivated chaos of rearranging DNA (and doing more cutting and pasting than a kids' TV presenter) the immune system manages to ensure lots of variation in the antibodies our B cells produce.

But, we're still not finished yet. The B cells have one last technique to send antibody diversity soaring to astronomical heights, and it has a name worthy of any space programme: somatic hypermutation. This begins once the V(D)J gene segments have been arranged in a satisfactory manner. The B cells display thousands of copies of their newly minted antibodies on their surface, looking like microscopic tennis balls covered in Y-shaped matchsticks. The naive B cells (naive because they have yet to meet a foreign antigen) leave the bone marrow and travel round the bloodstream. Chemical messengers beckon these B cells to our lymph nodes where they can be presented with foreign antigens such as chewed-up bits of bacteria. For many of the B cells the presented antigen

---

* Self-cells are exactly what they say on the tin – our body's own cells.

won't be of any interest because it won't fit their antibody, so they'll exit the lymph node and return to meandering round the blood some more. However for those B cells capable of binding to the antigen on display, as soon as they connect with it a cascade of events kicks off as quick and definitive as kissing a frog prince.

First, the B cells begin to replicate and transform themselves from naive B cells into plasma cells, which are a type of white blood cell and the antibody-producing powerhouses of the immune system, spitting antibodies out at a rate of 2,000 molecules a second. However, this process is about far more than simple replication: it's about making a mutant army. Each newly created cell is a little mutant copy of its parent, producing a totally unique twist on the antibody its parent produced. The secret ingredient in this feat of mutant-making is an enzyme called activation induced deaminase (AID). AID skips its way through the new B cell's DNA to the regions responsible for the antibody heavy and light chain formation. It then changes the structure of the DNA by altering cytosine molecules. Cytosine is one of the four 'bases' or building blocks of DNA (cytosine, guanine, adenine and thymine). If you picture DNA as being a bit like a zip, the bases are the teeth that stick out and connect with the opposite side. The four bases of DNA are paired so that cytosine on one strand always matches with guanine on the other strand, and adenine always pairs with thymine. This A:T G:C pairing is fundamental to how DNA operates. It's the linchpin of the whole system of inheritance and it's what AID seeks to mess with.

AID transforms the cytosines into a substance called uracil, which is a building block of a molecule called RNA but is definitely not meant to be found in human DNA. The presence of this misplaced molecule triggers the activation of DNA-repair enzymes, which work to excise the uracil and reinstate the cytosines. However this is an error-prone process and mutations arise as additional base pairs are accidentally added or deleted. The result is a mutation rate a million times higher than in other bits of our DNA, and this is responsible

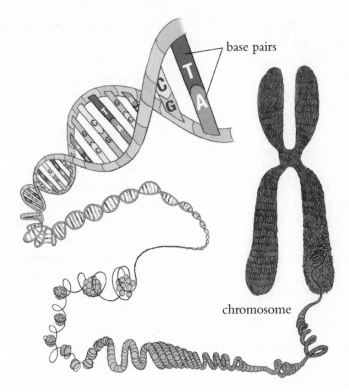

Figure 9.2 Base pairs.

for our stellar diversity of antibodies. Many of these antibodies will be unable to bind to the originally presented target antigen, or will do so only weakly, and the B cells that make those particular antibodies will die. However a handful of the newly minted B cells will create antibodies that are able to bond to the antigen better than those produced by the original parent B cell. These B cells will be stimulated not only to survive but to thrive and multiply. The cells created from this cell division will then undergo somatic hypermutation and create a clutch of even better targeted antibodies. Through these repeated B cell cycles of multiply and mutate, the body hones the antibodies it's producing to form progressively stronger bonds to the offending antigen, becoming better equipped to annihilate it.

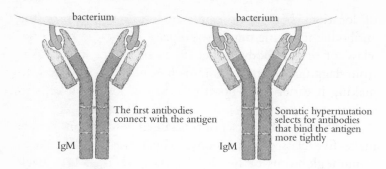

Figure 9.3 Somatic hypermutation.

A sting in the tail

This brings us to another question: how does a microscopic
Y-shaped protein annihilate anything? So far we've focused
on the arms of the antibody, but the answer to this quandary
lies in its tail end. Initially the antibody is stuck in the outer
membrane of the B cell that makes it, where it forms the
main part of what's called the B cell receptor. About 100,000
B cell receptors are studded across the cell surface, with the
arms of the antibody poking outwards to expose the antigen-
binding sites so they can stick to invaders like bacteria and
viruses. However the structure of the B cell receptor relies
on the tail end of the antibody staying anchored inside the
B cell, which is problematic because this is the bit that
actually enables the antibody to damage things. It's not as
epic a design fail as it sounds. As we said earlier, once the B
cell finds an antigen its receptor can bind, like Cinderella
fitting her foot into her glass slipper, the B cell is transformed
and, slightly less like a Disney princess, becomes a plasma
cell capable of firing out antibodies into its surroundings.
Though if anyone is listening, *that* is an animated film I'd
like to see. Ideally with a great musical number like 'Do
You Want to Build a B Cell' or 'The Circle of Somatic
Hypermutation'. Ahem. As we noted earlier, the B cell has
the capacity to make just one shape of antigen-binding site
its whole life, but it can cause a variety of effects in a variety

of locations because it can switch the tailpiece of the antibodies it's firing out. The tailpiece of choice defines the 'class' of the antibody, which in turn defines its effects – from flagging a microbe as munchable for macrophages, to making it easier for a flood of other white blood cells to arrive on the scene.

As we said at the outset of this tale of destruction, people make five classes of antibody, which are also known as immunoglobulins or Igs: IgD, IgA, IgM, IgG, IgE. Each class has its own modus operandi for attacking microbes and together they form a fighting force that rivals Marvel Comics' Fantastic Four. IgM would be the Thing. It's BIG. IgM is unique among the antibody classes because it's secreted from the B cell as a pentamer, which means five IgM molecules are bound together to form a sort of massive antibody throwing-star with 10 binding sites. If just one of these binding sites connects with an antigen, the complement cascade is triggered with all of the bacteria-bashing brilliance we discussed earlier. IgM is also noteworthy for being the first antibody all B cells produce and the most important in the early stage of an encounter with a microbe we haven't met before. This means doctors can use the presence of high volumes of IgM in someone's blood as a clue that they have a current or recent re-infection with a disease.

It's not just doctors that can use IgM to investigate disease; it's also a vital piece of evidence for scientists trying to solve a medical mystery: how can we tell how many people in a population have had an infectious disease if the people themselves don't know? For instance, have you ever had a sexually transmitted infection (STI)? No? Wrong. Well, probably. If you're sexually active, statistically speaking you've probably been infected with Human Papilloma Virus (HPV) at some point in your life. HPV is the most common STI known to man (or woman) and while it can cause genital warts, it's usually asymptomatic. Moreover 90 per cent of people who catch the virus clear it themselves within two

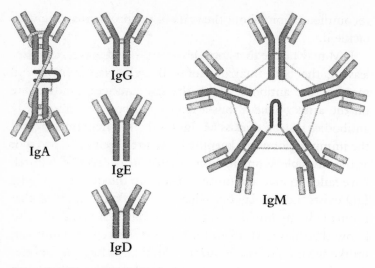

Figure 9.4 Five antibody classes.

years, which is why we can be walking around completely unaware of the residents of our nether regions. But if you don't know about it, how can I be reasonably sure you've had an STI without even meeting you? The answer lies in antibodies. Anyone who has been infected with HPV will have HPV IgG or IgM antibodies in their blood. Scientists can use this fact to track infectious diseases by conducting seroprevalence studies,* which involve testing blood samples from large numbers of people for these antibodies. These samples come from a range of sources, from antenatal clinics to blood banks, but they are entirely anonymous so the person the blood belongs to has no idea about the result. This raises ethical quandaries, because the irreversible anonymisation processes mean researchers cannot contact individuals and let them know the results. However it does allow researchers to calculate a good estimate of how common a disease is in a population and therefore make

---

* Seroprevalence is the occurrence of a disease or condition within a population, as measured by blood tests.

recommendations about the scale of the problem and how to tackle it.

IgM may be the first antibody every B cell makes, but after leaving the bone marrow most B cells 'class switch' and manufacture antibodies with the same antigen-binding site on the arms of the Y but a new tailpiece, creating class D antibodies instead of class M. IgD is the most enigmatic of all the immunoglobulins, keeping its nature about as easy to spot as the Invisible Woman. More than four decades of research have failed to give a definitive answer to the question of why IgD exists. Let me lay out what we do know: we know that it's made by mammals and bony fish, but not birds. We also know that like IgM it can form a B cell receptor. However, unlike IgM, it's released into the blood in such tiny quantities that it's unlikely to play a significant role in defence when roaming the blood by itself. It's also been established that people who are unable to make IgD don't appear to be any more susceptible to infection than the rest of the population. Listing these clues feels a bit like being Poirot at the end of a mystery. Unfortunately dear reader, unlike Poirot, I lack the big dramatic reveal. All we have is an inkling that IgD plays a part in modulating the immune response, but the hows, whats and whys of that theory are still full of mystery. This puzzle was perhaps best summed up by a 2006 article in the journal *Immunology*, which had a pretty good go at exploring the mystery before closing with a recommendation to lose 'the view that IgD has a simply definable function'. Yes, probably best for now. Given that IgD seems to have evolved over 470 million years ago, and we're still making it, it's probably safe to assume IgD has an important role, we just haven't discovered it. Yet.

While IgD has science all at sea, IgG takes us back to the knowledge equivalent of dry land because we know so much more about it. Arguably it's the Mr Fantastic of the Igs – leading the pack as the major antibody in the blood, representing about 73 per cent of all the immunoglobulins. One thing that makes IgG unique is that it's the only antibody that can cross the placenta from mum to baby during pregnancy. Unlike the pentameric groups of IgM, IgG excels by itself, with each antibody molecule released as a single soldier. In terms of skills, IgG can activate the complement cascade we met in Chapter 3, and like complement it can mark targets as munchable. Thus any bacterium, virus or parasite coated in IgG finds itself the yummiest dessert on the buffet cart and every hungry macrophage rushes to get itself a tasty treat.

IgE is the Human Torch of the antibody classes, with the capacity to turn your skin bright red and set it on fire with an incredible itch. It's barely there in the bloodstream, making up a tiny fraction of all Igs, but it's the main player when it comes to allergy and anaphylaxis. This is because IgE does a great job of stimulating a type of white blood cell called a mast cell that lives inside our skin, gut and other tissues. Every mast cell is packed with granules full of chemicals including histamine, which works on blood vessels, making them dilate wide open and causing their walls to become leaky. This leaky state makes it easier for lots of other white blood cells to exit the bloodstream and mount an attack against the intruder. However, if the IgE is responding to something harmless (as is the case in an allergy) it's positively irritating and potentially lethal. We can see this in the skin of people who have an allergic reaction to something they touch – their skin comes up in red, swollen, itchy blotches. Which is irritating, but much less problematic than anaphylaxis, where the reaction goes into overdrive and can cause the airways to swell shut. This is just a sneak peek into the world of allergy, which we'll have a rip-roaring tour of later in the book.

So, the numerate among you may have noticed my Fantastic Four analogy is about to encounter a mathematical challenge as we hit the fifth class of antibody. Awkward. Moving swiftly on, the interesting thing about IgA is location, location, location! It has the rather unglamorous title of the 'antibody of secretions' – snot, saliva, breastmilk and tears are prime real estate for IgA. To survive in these environments IgA has adaptations unique among the antibody classes. It is secreted as a dimer, that is a pair of antibody molecules bound together, an arrangement that's vital for shielding the IgA from enzymes in the secretions, which would otherwise digest and destroy it. Given IgA's key role in secretions it's unsurprising that one of its key killer skills is to block any wannabe invaders from making their way inside us. IgA latches on to specific sites on the surface of the bacterium or virus that they need to gain entry to our cells. By attaching at these points, the IgA effectively blocks and disables the invaders' docking stations, stopping them from latching on and entering our cells.

Let's talk about T cells

B cells and their multi-class antibody armoury have the ability to launch a tailored assassination campaign against almost anything, but they don't do it alone. They have a partner in crime in the shape of another type of white blood cell called the T cell. T cells complement B cells because they do something B cells can't: they can find and kill invaders that have made it *inside* our cells.

T cells get their name from a peculiar gland in the chest called the thymus gland where the T cells mature. The thymus is an odd little body part, whose name has been said to derive either from its similarity to the herb thyme or from a Greek word meaning heart because it's located near the heart. What makes it a little weird is that its size peaks in puberty, then gets smaller as we get bigger. So if you're an adult reading this, by now your thymus is a shrivelled shadow of its former self, with much of the original tissue replaced by fat. This sounds unpleasant, but if the thymus gland were to be removed at this point it would have little impact on you because it's effectively already done its job. However if the thymus gland is extracted from a newborn it has a devastating effect on the immune system, decimating the number of T cells in the blood and causing a slow, fatal wasting disease.

Thankfully, disorders of the thymus are exceedingly rare and a healthy thymus matures several types of T cell within it. The most notable are the helper T cell and the cytotoxic T cell, which take different approaches to achieving the same aim: the exquisitely tailored assassination of bacteria, viruses and anything else that dares enter the body. Cytotoxic T cells take a direct approach, killing any of our own cells that are harbouring a pathogen such as a virus. The helper T cells do what it says on the tin, 'helping' other immune cells such as B cells, cytotoxic T cells and macrophages in their quest to kill invaders. The activation of either of these abilities occurs once the T cell has met its matching target, just like when a B cell finds its complementary antigen. However both types of T cell interact with antigens in a different way from B cells. While the B cell can bind to invaders roaming free in the body, T cells can only recognise an enemy that's been chewed up and presented to them by an infected host cell or a 'professional' Antigen Presenting Cell (such as a macrophage or specialised B cell).

T cells interact with their quarry through a T cell receptor, which pokes out of the cell like the eye-piece on a Dalek. The shape of the T cell receptor has been brought to light by an imaging approach called X-ray crystallography, which is a nifty technique that showcases the dark art of physics. An X-ray is beamed at the receptor and by studying how the beam spreads out and shifts its angles and intensity after hitting it, we can build up an image of the receptor – a bit like looking at the shadow cast by an object. In the case of the T cell receptor its structure is a lot like one arm of the Y that makes up an antibody. Like antibodies, T cell receptors are made from a pool of V, D and J gene segments, which are put on shuffle to create a multitude of different T cell receptors. However, unlike antibodies they don't undergo cycles of somatic hypermutation to create T cell receptors of increasing specificity to an antigen. This means T cells don't get the chance to improve on their initial relatively weak ability to bond to the target, so a few extra receptors are needed to kick-start the T cell response.

These different receptors are essential to a highly complex secret handshake that occurs between the T cell and the cell presenting the mushed-up microbe to it. In the case of helper

T cells and professional Antigen Presenting Cells, one party has essentially spat on its hand as the APC's extended receptor includes a gob of antigen the cell has chewed up internally before proffering it to the T cell. The second element of the handshake acts to bolster this grip as additional receptors on the T cell hook up with their dancing partners on the surface of the Antigen Presenting Cell and add a frisson of encouragement to the T cell, spurring it on to activation. Cementing the entire interaction are adhesion molecules present on the surface of both cells. These molecules lock on to each other like glue and hold the APC-to-T cell handshake firmly together for long enough for activation to occur.

The secrets of this handshake can be subverted by those in the know, allowing us to engineer formidable weapons to unleash on previously untreatable diseases. For instance, the use of genetically modified T cell receptors is currently re-writing our approach to treating leukaemia in children and adults who would otherwise have few, if any, treatment options. The patients' own T cells are harvested from their blood and genetically modified so they display lab-created chimer antigen receptors (CARs) on their surface that are specially designed to bind to the leukaemia cells. The CAR T cells are then grown by the billion in the lab before being returned to the person's bloodstream where they set about identifying and destroying the leukaemia cells. It's worth emphasising that this technique is still in the early stages of development, with much to be learned and challenges to be overcome. One such challenge is that trials have found that the CAR T cells can over-activate the immune response, which then needs to be suppressed with drugs – a delicate balancing game to play with very sick people. Nevertheless the early results are promising, with a clinical trial published in 2014 finding that treatment with CAR T cells led to all signs of cancer disappearing in 27 out of 30 patients. Of these, 19 achieved sustained remission, 15 of whom had no further treatment. This is a significant survival improvement on what doctors had ever previously dared to hope for these patients, and researchers in the field think CAR T cells could act as a bridge to get patients into a position to receive bone marrow transplants when chemotherapy fails. Other small trials including different types of leukaemia have also been successful, demonstrating that CAR T cells have the potential to create options for those who previously had none.

CAR T cells aren't the only way we can manipulate the adaptive assassins to our benefit. One of the seminal victories in our war on bugs was the day vaccination came to be, allowing us to treat, prevent and even exterminate some of the most deadly diseases on the planet. In our next chapter we'll travel back to the beginnings of this story, to a time when smallpox was rife, modern vaccination wasn't even a twinkle in the eye of its creator and the wife of the British ambassador to the Ottoman Empire was injecting public health principles into the aristocracy, once it had been tested on orphans and convicts. Naturally. Onwards to Chapter 10, Jeeves ...

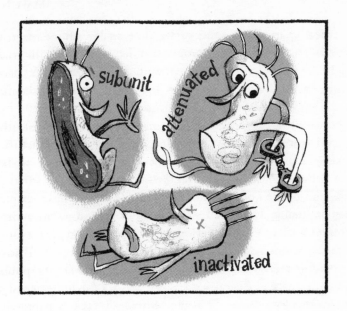

CHAPTER TEN

Vaccines: A Triumph of Man's Manipulation of the Immune System

Lady Mary Wortley Montagu is an unlikely heroine of public health. Born into money and beautiful to boot, she lived the lavish lifestyle of an English aristocrat, shielded from the squalid slums and sewage-spiked drinking water inflicted on London's poor. However, not every scourge in the eighteenth century could be side-stepped by the upper echelons of society. In 1715, at the age of 26, Lady Mary was lucky to survive an encounter with the brutally egalitarian smallpox. While she escaped with her life, the scarring from the skin eruptions left her pretty face terribly disfigured.

Tragically, smallpox marked her life again just 18 months later when it killed her younger brother.

These experiences travelled with our heroine when she journeyed to Constantinople in 1717 with her husband, who had been appointed as an ambassador to the Ottoman Empire. There she encountered the local practice of variolation, which aimed to protect people from smallpox (Variola) by inoculating them with a small quantity of the virus. A fresh pustule on the skin of an infected patient would be lanced and then a soupçon of the goopy contents needled under the skin of the person being variolated. The idea was that exposure to smallpox through variolation would protect against a full-blown smallpox infection later in life.

Having seen what smallpox could do, Lady Mary was most taken with the concept and wrote to her friend about this interesting foreign fad. She then went further and commanded the embassy surgeon, Charles Maitland, to variolate her five-year-old son. Lady Mary was so pleased with the results of the procedure that when she returned to London with her family in 1721 she had Maitland variolate her four-year-old daughter in front of the assembled physicians of the Royal Court. Through this act of health promotion she kick-started a series of events that would save many lives and act as a precursor to modern-day vaccination.

Maitland and Mary's little performance piqued the interest of members of the Royal Family, but they required more assurances before having their own little darlings variolated. So Maitland was told to set up a trial with six prisoners who were promised the King's favour in return for being variolated. Thankfully they all survived to enjoy the King's favour, and were immune when subsequently exposed to smallpox as a test. The great and good, who had been eyeing the trial like careful spiders, pounced on this success and declared the next group of 'volunteers': orphans. Only after the orphaned children had survived the same trial was Maitland allowed to variolate the two daughters of the Prince of Wales. Just as today, gorgeous Prince George can make a cute blue cardigan a must-have sell-out, so royal approval in the eighteenth century made variolation *de rigueur*. From pampered princesses to the seasoned soldiers of King Frederick II of Prussia, variolation swept through Europe.

However, unlike little blue cardigans, variolation could be deadly. Two to three per cent of variolated people died from smallpox, or caught another disease from the procedure (such as syphilis), or became the starting point for a further smallpox epidemic. This raised questions about whether variolation was better than the harm it was designed to prevent. To answer such questions Dr Zabdiel Boylston and Reverend Cotton Mather studied Boston's great smallpox epidemic of 1721, which saw 50 per cent of the 12,000-strong population infected. In one of the earliest applications of medical statistics the doctor and the good reverend compared the mortality rate of variolated and non-variolated individuals, and found that while 12 per cent of non-variolated sufferers had died, only 4 per cent of the variolated died. These rather more favourable odds vindicated the beleaguered Boylston and Mather, who had suffered multiple threats and one house-bombing courtesy of violent anti-variolators.

Variolation continued to be practised for many years in Europe and the US, though evidence exists that inoculation with smallpox was old hat in China, where it had been happening for many centuries before catching on in the West. However, towards the end of the eighteenth century there was something new and shiny coming to the world, something that would tear open a rich new vein of opportunity to push back against the flow of infectious diseases spilling over the planet. The something was vaccination and its story begins with a boy called Edward.

Edward Jenner was variolated against smallpox as a child, but as a man he would be credited with the scientific validation of a safer approach to preventing smallpox, a little something called vaccination. Unlike Lady Mary, Jenner was a very likely public health hero – he was a doctor who had laboured under the tutelage of John Hunter, one of the most famous surgeons in England. He also had the sort of extracurriculars on his CV that would make any regular human feel slightly inadequate. These included classifying some of the species brought back from Captain Cook's first voyage to Australia (he turned down the opportunity to sail on the second trip) and being the first person to discover that cuckoo chicks kick their foster siblings out of the nest. However, it was his work on the theory that smallpox could be prevented by exposure to the milder disease cowpox that made him a namesake of lecture theatres and academic institutes, and a general public health luminary.

Unfortunately his stellar work began by carrying on the dubious tradition of practising on children. He harvested material from a juicy cowpox pustule on the arm of a dairymaid and inserted it under the skin of an eight-year-old by the name of James Phipps. After a minor feverish illness, the boy was ready for Jenner's next step: introducing some material from a smallpox lesion into the boy's skinny arm. Thankfully Phipps remained disease-free, and clutching his proof-of-concept Jenner set about collecting more evidence to demonstrate that vaccination with cowpox was the way to fight smallpox. Others had made this connection before Jenner, but it was the way he collected a body of evidence that meant his was the name that made it into the history books as the father of modern vaccination.

While variolation involved exposure to smallpox, vaccination only involved exposure to an agent incapable of causing smallpox. This meant vaccination caught on rapidly, aided once again by orphans and royalty. King Charles IV of Spain, having lost family members to the disease, announced in 1800 that the vaccine would be available to all of his subjects. He wanted to extend this to Spain's South American colonies, but faced the challenge of transporting the vaccine long distances without a fridge. Cue orphans. They put 22 orphans on the boat, and used them like a living chain storage system. The first was given the vaccine, and when they developed wet skin lesions a few days later these were lanced and the ulcerous material passed into the next child, and so on. So when the mission arrived at its destination, they still had a supply of cowpox alive and well in the pustules of the final child.* By the time they were finished the expedition had vaccinated thousands of people across multiple countries.

The work started centuries ago culminated in the official eradication of smallpox in 1980, and today the virus lives in just two high-security laboratories, one in Russia and one in the US.† Arguably the smallpox vaccine is therefore no longer required, however, ongoing bioterrorism fears mean that the US has enough stockpiled to vaccinate every man, woman and child in the country. Similarly, the UK government spent an undisclosed sum to buy enough vaccine to cover half of the UK population in the wake of 9/11.

A one-man vaccine machine

Terrorism aside, vaccines exist to protect people from naturally occurring infections that present a very real and present danger to human health. Today we have vaccines for a diverse range of diseases, from chickenpox to cholera, measles to mumps, rabies to rotavirus. Incredibly, we have just one eccentric (to put it mildly) but brilliant man to thank for more than 40 of these vaccines. Maurice Hilleman invented vaccines to prevent measles, mumps, rubella, hepatitis A, hepatitis B, chickenpox and more, with estimates that his life's work could have saved up to eight million lives. He also discovered the adenoviruses, a family of viruses that commonly cause cold-like symptoms, and was the first person to purify interferon, an immune messenger chemical now used to treat hepatitis B, multiple

---

* While this wouldn't make it past an ethics committee today, the story did at least have a faintly happy ending for the children, who were rewarded by being adopted and educated in Mexico courtesy of the Spanish government.

† And occasionally in a not-so-high-security cardboard box. In 2014 vials of smallpox thought to date to the 1950s were found in a cardboard box in a disused storage room belonging to the Food and Drug Administration. Despite their being vintage, subsequent tests found that some of the samples were still capable of growing. 'Oops' doesn't really cover it.

sclerosis and some cancers. He worked seven days a week, something which contributed to one of his greatest challenges: finding a suitable wife. It's tough when, as he put it, 'You don't know whether they're drunkards, or they'll spend all your money, or whether they have venereal diseases.' Given this quote, I can't help but feel Maurice was looking for love in all the wrong places. However, his next plan wasn't particularly appropriate either. He asked his assistant to go through the CVs of women who had applied for jobs at Merck, the pharmaceutical company he was working at, and pick out some likely candidates for a very different kind of engagement from the one they had applied for. The plan was to do that once a week until they found Hilleman the right woman. Now there's a story to tell the grandchildren. Eventually Maurice found his leading lady and had children, one of whom caught the mumps virus, which he hastily cultured and used to make the mumps vaccine.

Part of what renders Maurice Hilleman's achievements so impressive is that making a vaccine that can trigger a lasting immune response without causing serious side-effects is easier said than done. There are six known strategies for achieving this goal, each of which has its own pros and cons: live, killed, subunit, conjugate, toxoid and recombinant vaccines.

Vaccine type 1: live vaccines

Option one is to take a living microbe and weaken it in a lab to create a live, attenuated [*i.e.* weakened] vaccine. It's a bit like taking the microbe, tying its hands behind its back, then setting the immune system on it. In some ways this is the grand-daddy of vaccines because the immune system is taking on a vulnerable version of the real pathogen, which is an ideal way to train the body's defences. The result is a strong immune response, and a lasting memory of the encounter which means just one or two doses of the vaccine can bring about lifelong protection.

There are downsides too of course. The main one is that the microbe's hands don't necessarily remain tied for ever – as it multiplies and mutates there is the remote possibility that the microbe becomes capable of causing the disease again. Partly because of this, people with weakened immune systems, due to HIV or leukaemia for instance, can't be given live vaccines because there is a risk that their system doesn't get on top of the vaccine and it makes them ill.

Another challenge, as King Charles IV well understood, is that transporting live microbes is a delicate process. This is especially true in low- and middle-income countries, which may lack the robust refrigeration systems necessary to keep the vaccine cool and alive until it's ready to be administered. One company that has cracked the cold supply chain in almost every nation on the planet is Coca-Cola, and they've been helping health systems in some countries to overcome the chilled vaccine problem. This is a mutually beneficial arrangement, giving Coca-Cola some very positive PR, while helping countries like Ghana understand how different incentives and processes can keep the cold chain intact for vaccines and beverages alike.

The MMR scandal

One of the most infamous live, attenuated vaccines is the measles, mumps and rubella combined vaccination (MMR). MMR hit the headlines in the British press in 1998 with suggestions of a possible link between the vaccine and autism, though with less subtlety than that. A selection of headlines from a British tabloid, the Daily Mail, gives a good snapshot of the coverage: 'MMR killed my daughter', 'MMR safe? Baloney. This is one scandal that's getting worse'. The source of the furore was a research paper in the high-class academic journal the Lancet which described 12 developmentally challenged children, and stated that: 'Onset of behavioural symptoms was associated by the parents with measles, mumps, and rubella vaccination in eight of the 12 children.' One of the lead authors of the paper was Dr Andrew Wakefield, then a researcher and senior clinician at the Royal Free Hospital in London, now an aspiring reality TV producer. But we'll get to that.

At the time, Wakefield's study of a handful of kids, his publicity tour de force and the media acting as a willing platform all combined to shatter public confidence in the MMR and started a sharp decline in the use of the life-saving vaccine. In 1996/97, 92 per cent of English children received the MMR, but by the end of 1998 this had dropped to 88 per cent. MMR's fall from grace continued until 2003/04 when the vaccination rate in England bottomed out at 79.9 per cent, far below the World Health Organisation's recommended 95 per cent minimum. In some ways MMR was a victim of its own success, as many parents had forgotten the severity of the diseases it had been designed to prevent. Measles can cause lethal pneumonia, mumps can infect the brain and if rubella is passed to a pregnant woman it can make the baby deaf, blind and stop the heart and brain developing normally. People became ill who probably would not have if the MMR scare had not happened.

And it was utterly needless. The *British Medical Journal* (BMJ) has since called the study an 'elaborate fraud'. An article in the BMJ by award-winning investigative journalist Brian Deer painstakingly detailed how Wakefield was being paid by a solicitor for two years prior to the unveiling of the science-juddering publication. This payment was to support a lawsuit that required evidence of a bowel-brain syndrome of exactly the type Wakefield et al 'discovered'. All in all Wakefield took home £435,643 ($575,005) in confidential payments. Deer also stated that the article's descriptions of the children's problems were inaccurate and that the claim that the onset of autism corresponded closely with receiving the MMR were just plain wrong. Moreover, it was reported that eight months prior to the paper's publication Wakefield had filed a patent for a novel measles vaccine to replace attenuated viral vaccines like MMR.

The General Medical Council (GMC), the body that sets standards for doctors in the UK, spent two and a half years investigating Wakefield and in 2010 concluded that he was 'dishonest, irresponsible and showed callous disregard for the distress and pain' of children. His failure to disclose the conflict of interest posed by the lawsuit led the *Lancet* to question the paper's 'suitability, credibility, and validity for publication'. Consequently the *Lancet* retracted the paper and the GMC banned Wakefield from practising medicine in the UK. Eleven of the thirteen authors on the paper withdrew their support for the article. Wakefield, however, did not. He went to the US, has written a book on the subject and in 2013 could be found pitching a reality series on autism to TV execs. He maintains that the research was not fraudulent.

Why rumours matter

As for MMR, large, robust and undisputedly fraud-free trials have found zero evidence that it causes autism. However, when it comes to trust in vaccines, evidence isn't necessarily enough. The Vaccine Confidence Project exists to monitor public confidence in immunisation projects, using a variety of approaches including combing through social media platforms to see what's being said about a vaccine in a particular country. Using this technique they can give a government an indication of how positively or negatively their programme is viewed. They can also trace the spread of rumours about vaccines, which can see stories bouncing back and forth across the globe and affecting vaccine uptake worldwide. One example is the tetanus vaccine in Kenya, where a group of Catholic bishops issued a statement in 2014 raising multiple concerns about the vaccine, including a revival of a 20-year-old myth that it was a form of stealth sterilisation by wealthier countries. The original myth had spread from the Philippines to Mexico, and then to Bolivia and Nicaragua. Tetanus vaccination uptake dropped by a staggering 45 per cent in the Philippines as a result of Catholic organisations getting behind the sterilisation idea. It got worse. The vaccine stopped being administered at all in Nicaragua, and Pro-Life groups in Mexico persuaded several national legislators to support charging the Secretary of Health with genocide. In an attempt to revive confidence in the vaccine, the WHO had it tested at independent labs, which found no evidence to support the theory. That wasn't enough to convert the church, so they let the Vatican test the vaccine at a facility of their choice. The resulting negative test did eventually appease the church leaders, who were the real guardians of public opinion, not the WHO and not vaccine scientists.

One of the reasons public consensus matters is that many immunisations for contagious diseases like measles, influenza and rotavirus work part of their magic via 'herd immunity'. Essentially, if enough people are immunised the pool of people left for the bacteria or virus to infect is tiny. This makes it hard for an outbreak of a disease to occur, because the microbe can't find enough people to infect and so it quickly dies out. Therefore when enough people are vaccinated, there is a degree of protection even for those who can't be vaccinated, like newborns, pregnant women or the very ill. The threshold for herd immunity to work is high – in the case of measles, 19 out of 20 people need to be vaccinated to achieve herd immunity. When too few people are vaccinated, for instance in the MMR scare, the herd immunity fails and the vulnerable don't enjoy the same protections.

This can play out on a local level, so even if vaccination levels are high in the country as a whole, if a particular community has a low rate of vaccination it can experience an outbreak of a disease. Anti-vaccination sentiment tends to pool within like-minded people, so this can be a significant problem. One example occurred in 2013, in Swansea, Wales, which was a part of the UK that saw an even more dramatic drop in the uptake of the MMR vaccine than the rest of the nation. The unvaccinated population were like fish in a barrel for the measles virus. Over the course of eight months, over 1,200 people fell ill, 88 had to be hospitalised and one man died from measles. A massive MMR drive offered 35,000 vaccinations to the local people, with 95 per cent of those vaccinated in the programme having never previously received a single dose. Eventually the outbreak was brought under control, but only after a lot of suffering and almost half a million pounds of spending on the response.

Polio is another vaccine that knows all about image problems. Until 2012 it was the modern poster child of vaccination thanks to a 25-year programme to vaccinate billions of children, which had successfully put the world on the verge of polio eradication. However in 2014 the World Health Organisation called a global health emergency, the second time it has done so since the concept began in 2007, because the virus had begun to spread once more. There were several reasons for polio's resurgence, including political instability and the lack of infrastructure to carry out a concerted vaccination campaign in some countries. However, another factor was the decimation of trust in vaccination programmes which occurred after the CIA used a vaccine programme as a cover to assassinate Osama bin Laden in Pakistan. The CIA has since said that vaccination programmes will never again be used as subterfuge for spying, but the damage had been done. Militants in Pakistan targeted those involved in anti-polio campaigns, with 60 polio workers and security staff reported killed between December 2012 and April 2014. The price has been borne by the children of Pakistan, as the country remained one of only three nations on the planet where polio was still endemic in 2015.

Vaccine type 2: killed vaccines

As well as espionage, the history of the polio vaccine includes immunological intrigue with two very different ways of skinning the same cat. There is the oral polio vaccine (OPV), which is a live, attenuated vaccine like MMR, and the inactivated polio vaccine (IPV), which is an example of a completely different vaccine type known as a killed or inactivated vaccine. This type of vaccine is made by killing the bacteria or virus in a lab using chemicals, heat or radiation. As the microbe is dead there is no possibility of it mutating and causing polio; while the risk with the live vaccine is minuscule, a zero risk is preferable with a disease that has paralysed and killed thousands of children.

So if OPV is more dangerous, why do 150 countries still use it? They're choosing the best set of trade-offs for their population. For instance, most killed vaccines trigger a weaker immune response than live vaccines, which often means multiple booster shots are required to ensure continued protection. IPV, for example, fails to induce a strong immune response in the gut, which means that if someone catches the wild polio virus it won't make them ill but it can still multiply in their gut and pass out in their poo. This makes it less effective than OPV at stopping the spread of the disease within communities. However, of the two vaccines, IPV is the safer option because of the aforementioned zero risk that it will cause polio in the person vaccinated. Thus in countries where wild polio has disappeared, like the UK and the US, the safer killed vaccine is the standard choice. In time this will be the case for every country on the planet as part of the WHO's 'end game' to send polio to the eradicated disease list. It's only got smallpox on it at the moment, so apart from ending the death and human suffering of polio, it would also make the concept of an eradicated disease list a reality rather than an aspiration.

The inactivated polio vaccine may be part of the momentous eradication of polio one day, but the story of its development casts a shadow we should be equally mindful of. The shadow stretches from its catalyst the HeLa cell line, which is intertwined with some of science's greatest triumphs and darkest failings. The first time I encountered the story of HeLa cells – and Henrietta Lacks, who they are named after – was in a cell biology lecture during which I'd scribbled in the margins 'more of her cells on earth now than when she was alive?!' For decades the story of HeLa was told like an honour roll of scientific discoveries, but the story we should all sit up and listen to is that of Henrietta Lacks, the woman they were originally taken from. In Baltimore in 1951, Henrietta Lacks was 31 years old, a loving mother to five young children and sadly in possession of a body riddled with an aggressive cervical cancer that would kill her before the year came to a close. During her time in hospital, and unknown and unconsented to by Henrietta, her doctors took a sample of Henrietta's cervix – not for her benefit but for a researcher called Dr George Otto Gey.

Gey discovered something remarkable about the cancer cells: they were immortal. Until this discovery, medical research had always been held back by the fact that cells grown in labs always died eventually, regardless of how they were treated or prepared. The HeLa cells, on the other hand, were unstoppable. In 1952 the first 'HeLa factory' was set up as a non-profit organisation to make vast quantities of HeLa cells to be shipped to researchers in need. One scientist in need was Dr Jonas Salk, who used vats of the cells in the creation of the inactivated polio vaccine. Since then HeLa cells have been used in everything from cancer research to the creation of human–animal hybrid cells. It has been estimated that if you collected up all the HeLa cells ever grown, you'd have 50 million tons of cells on the scales. That's probably about a billion Henriettas. Yet it wasn't until 20 years after her death that her family were even aware Henrietta's cells were still alive, by which time the production and application of HeLa cells had become very much for profit. While Henrietta's children could not even afford medical care, they had to try and wrap their heads around the idea that their dead mother's cells were very much alive and generating many millions of dollars for complete strangers. Thanks in part to a book by Rebecca Skloot, which does a beautiful job of giving a voice to Henrietta and her family, the medical research community and the world at large is now far more aware of the role Henrietta Lacks played in science's successes. Modern codes of practice for ethical research should prevent the sort of violation of basic rights to self-determination and dignity the Lacks family endured, but we must also never forget that medical research is meant to serve and enrich our humanity, not denigrate it.

One vaccine that humanity is indebted to takes on a disease that has been with us for centuries and has slaughtered millions: influenza. One of the main reasons influenza is so deadly is that its surface appearance mutates rapidly, creating an ever-shifting target for our immune system to attack. This is also a challenge for vaccine creation because we can't develop a single flu vaccine that provides cover for multiple flu seasons. Instead there is an annual programme of vaccination with a cocktail of killed vaccines tailored to that specific year and hemisphere. The vaccine cocktail mixologist is the World Health Organisation, which decides upon the three or four most likely culprits for causing flu outbreaks in the coming year and advises vaccine manufacturers to make a vaccine to cover those strains. For instance in 2014/15 the Northern hemisphere flu vaccine included three types of virus: A/California/7/2009 (H1N1) pdm09-like virus, A/Texas/50/2012(H3N2)-like virus and B/Massachusetts/2/2012-like virus. Predicting the future isn't easy so the mix isn't always quite right, and over the winter of 2014/15 public health surveillance staff in the UK noticed there were more deaths from influenza than normal and concluded the vaccine wasn't working as well as hoped. Normally the flu vaccine is effective in about 50 per cent of people, but in 2014/15 it was effective in just 3 per cent. The source of the problem was the H3N2 strain, which had mutated significantly, resulting in a wild virus very different from the version included in the millions of doses of vaccine produced months before. The mismatch resulted in the worst British flu season in years.

Vaccine types 3–6: subunit, conjugate, toxoid and recombinant vaccines

Influenza, like many microbes, betrays its presence to the immune system through tell-tale antigenic markers on its surface. It's a bit like wearing a Red Sox shirt in the midst of a gathering of die-hard Yankees fans, or a Man City one in a crowd of Man U supporters (epic fail). A third form of vaccine design, called the subunit vaccine, takes advantage of this concept by creating vaccines made just from the bits of the microbe that incur the wrath of the immune reaction – the Sox logo or the pale blue shirt in our sports fan analogy. Subunit vaccines can include anything from a single antigen to over 20 antigens. The necessary microbial body parts can either be collected from a chemically dissected microbe or they can be manufactured in a lab from the genes that code for the markers. The big advantage of this technique is that it limits the vaccine to the bare microbial bones you need, which reduces the risk of a bad reaction to other extraneous parts in the vaccine.

The subunit vaccine has a spin-off called the conjugate vaccine, which is perfectly suited to tackling super-stealthy microbes. Some bacteria have a sugar-based cloaking device that makes it particularly hard for the immature immune system of young children to recognise them. Conjugate vaccines get round this by giving a dose of the cloak that has had a recognisable marker stitched on to it. A bit like throwing a handful of dust onto the invisibility cloak, this helps the body recognise the cloaked bacteria, and should it ever be infected with the disease the immune system will swiftly spot the intruder and neutralise it. This technique is used to great effect in key childhood immunisations including the Haemophilus influenzae type B (Hib), meningococcal and pneumococcal vaccines.

Diphtheria is another disease whose prevention is part of the standard childhood vaccination schedules of many countries, but has a distinctly non-child-friendly story in its past: once upon a time there was a horse called Jim, and his blood killed small children. Tragic, bizarre and factually accurate, this story took place in 1901 when horses were the main source of antitoxin to treat people with active diphtheria. Jim was a retired milk wagon horse in St Louis who had been inoculated with diphtheria toxin to stimulate his body to make antibodies against the toxin (also known as an antitoxin). In three years of service, Jim's blood had been used to produce over 34 litres of antitoxin. Sadly, Jim contracted tetanus and his contaminated blood serum was used to make a batch of antitoxin that killed 13 children. This terrible loss highlighted the complete absence of appropriate quality control measures in vaccine development and provided the impetus for the Biologics Control Act 1902. Today's diphtheria antitoxin is still based on horse serum, which means it comes with a higher risk of causing a severe allergic reaction than human antibody products. Consequently if people get diphtheria skin infections, where it's unlikely lots of toxin will be absorbed, it is actually safer *not* to give the antitoxin.

Of course it is even safer not to come down with diphtheria in the first place, which is where our fifth vaccine class comes in: toxoid vaccines. There are some diseases where the toxin produced by the bacteria[*] is the real danger, and this includes diphtheria and tetanus. *Corynebacterium diphtheriae* produces the diphtheria toxin that may affect the skin, but can also cause lethal swelling and a thick grey-white coating in the throat. Tetanus is at least as unpleasant, with the tetanus toxin emanating from bacterium *Clostridium tetani* reaching into nerves and causing massive muscle spasms, leading to lockjaw. The proportion of tetanus sufferers who are killed by this virus varies from 10 to 90 per cent, with the highest rates found among infants, the elderly and in places with fewer intensive care beds. While it can't transfer from person to person it can often set up camp in soil, making it difficult ever to eradicate it.

In both tetanus and diphtheria the target of vaccination is the toxin. A purified form of the toxin is steeped in formaldehyde, which stops it from causing harm but maintains its structure well enough so that the antibodies produced against it are capable of neutralising the real toxin. Thanks in large part to childhood vaccination programmes, cases of diphtheria and tetanus are exceedingly rare in places like the

---

[*] A toxin is a poison secreted by the bacteria. One of the most famous examples is botulinum toxin, which is produced by the bacterium *Clostridium botulinum* and is one of the most powerful toxins known to science. It causes botulism, which involves symptoms like slurring of speech and weakness and kills 5–10 per cent of those infected. However, medicine has found another use for botulinum toxin: the foreheads of the rich and wrinkly. It's the basis of botox, which for about £300 ($377) freezes the muscles of your face and makes it look smoother.

UK and the US, with just 20 cases in the UK from 2010 to 2015. However the importance of such programmes was underlined when a diphtheria epidemic swept the countries of the former Soviet Union from 1990 to 1998. A decline in the number of children being vaccinated led to an epidemic that saw 157,000 people infected with diphtheria and 5,000 deaths.

In the late 1970s, Swiss foxes would have had good reason to believe in a benevolent God. At this time in the wilds of Switzerland the fox equivalent of a Happy Meal (fishmeal, fat and paraffin) rained from the sky. However, as is often the case with gods, they were working in mysterious ways. The bounty from the sky, which was often dropped by light aircraft or helicopters over tricky terrain, was laced with an oral rabies vaccine as part of a programme to eradicate rabies in the fox population of mainland Europe. The programme began in Switzerland in 1978 before sweeping through many other countries including Germany, Finland, France and Belgium, and played a key part in those countries becoming free from terrestrial rabies.* The vaccine used in the programme was either a live, attenuated vaccine or an example of our sixth and final vaccine category: the recombinant vaccine. The recombinant vaccine essentially forces a harmless microbe to display the surface markers of the microbe you're trying to vaccinate against. So in the case of the recombinant rabies vaccine, another type of virus called vaccinia virus is rendered harmless and then made to express rabies surface markers. It's a bit like making vaccinia wear a very realistic rabies mask, and it does an excellent job of teaching the immune system what rabies looks like with zero risk of actually causing rabies, which is good for foxes and people alike.

---

* Non-terrestrial rabies is a harder nut to crack; bats are the key carrier and lacing live bugs with vaccine for bats to eat is, unsurprisingly, difficult.

The benefits of blue blood

All vaccines, no matter what type they are, must be free from contamination. This means they should only contain the bits of bacteria or virus needed for the vaccine to be effective without any other interlopers being included – the earlier story of Jim and tetanus shows why. Today one way of checking whether vaccines and other drugs are safe from bacterial contaminants depends, bizarrely, on a crab's blue blood. It's a beautiful baby-blue, and in laboratories across the US right now scientists are harvesting the blood in bucket-loads from live crabs. However, crab is a misleading term for these 46cm (18in) long, 13-eyed wonders, which are more closely related to spiders and scorpions. They are so old that it's thought their first appearance on Earth pre-dates the dinosaurs by 100 million years. Our vampiric approach to these ancient armoured arachnids dates back to the 1950s when we discovered their blood contains a chemical that can detect tiny traces of bacteria and trap them in a tough clot-like mesh. The chemical is called coagulogen, and it's thought it evolved to help the horseshoe crab defend itself against the bacteria that live in abundance in the shallow coastal water the crab calls home. Unlike you and me, with our complex internal structure where bacteria have to worm their way through to get from our guts or lungs into our bloodstream, the horseshoe crab is structured a bit like a spider on the inside with very few separate compartments. As a result the horseshoe crab's blood sloshes around inside it, doing a beautiful job of bathing the crab's cells in nutrients, but simultaneously making it easy for bacteria to spread far and wide around the crab's innards. Given this open-plan approach to insides, the crab's saving grace is coagulogen's ability to ensnare bacteria in a web of impenetrable goop and thus prevent them from moving any further.

Unfortunately, as well as being a fantastic defence, coagulogen puts a price tag on the crab's head to the tune of $15,000 per quart (£11,364 per 1.1 litres) of coagulogen. This is because the pharmaceutical industry can use coagulogen to detect bacterial contamination in any solution they want to test. So if you've ever had a vaccination, you can thank the horseshoe crab for ensuring it was safe. Coagulogen is so sensitive that a concentration of bacterial toxin of just one part in a trillion is enough to activate its remarkable abilities and transform the contaminated solution into a gel. The test is known as the LAL (Limulus amebocyte lysate) test and its incredible ease, speed and accuracy has made it exceedingly popular with the pharmaceutical industry. Which is why companies collect over half a million crabs from the shallow waters off the east coast of the US, poke holes near their hearts and drain their blood to collect the coagulogen-producing cells.

Mindful of the need for a sustainable resource, they don't entirely exsanguinate the crabs, taking up to 30 per cent of each animal's blood before returning them to the sea a long way from where they were collected. I'd like to be able to tell you the crabs swim off into the sunset and live happily ever after. But I can't. The crab is definitely getting a raw deal from this arrangement – at least when we give blood we get a biscuit, a drink and the warm glow of having helped someone. The horseshoe crab gets a 10–30 per cent chance of death in the short term and, recent research suggests, a reduced fertility rate for females in the longer term. A team of scientists tracked some released female crabs after their blood-letting and found they moved more slowly than their non-harvested counterparts and were less likely to follow the tides – an important skill for breeding horseshoe crabs. This has led to suggestions that the harvesting process is responsible for a recent decline in horseshoe crab numbers.

However there may be a happy ending on the horizon as this decline (and the general faff of catching crabs) is spurring people on to create a synthetic alternative to horseshoe crab coagulogen. If we succeed, the crabs would follow in the footsteps of a long list of animals that have been happily usurped. For instance, until the 1980s diabetics relied on purified pig pancreases for insulin. Now we can take the gene for human insulin, pop it into *E. coli* and use the bacteria like little factories to produce massive volumes of insulin. The quest for synthetic coagulogen isn't quite that far forward, but we're nearly there, so the end may be in sight for our beleaguered blood benefactors.

Checking for contamination is just one step of many that help ensure vaccination in the twenty-first century is an exceedingly safe experience. One of the few serious risks is an allergic reaction, which although very rare, can be life-threatening. Sometimes we can predict who will have an allergic reaction; for instance, some vaccines use eggs in their manufacture and therefore people with an egg allergy should avoid them. Most allergic people can live a vaccine-avoiding life without too much danger as long as vaccination has a high uptake rate in the population. However this is not the case for all allergies. What if you were allergic to strawberries or peanuts or the cold? What if you were a gymnast allergic to exercise? Our next chapter has everything you're itching to know.

CHAPTER ELEVEN

Allergies: Everything You're Itching to Know

What does a kiwi fruit have in common with a condom? In what circumstances is stale pancake mix deadly? And why might a tick from Texas stop you enjoying a tasty steak? The answer to all of these bizarre questions lies in the itchy, scratchy land of allergies. An allergy is essentially an inappropriate immune response to an otherwise harmless substance (an allergen), be it pollen, prawns or penicillin. The concept of allergy in the scientific literature is a little over 100 years old, but the phenomenon is likely to be as old as we are. Being a science geek, and more importantly a pedant, I tried to find the oldest recorded example of an allergic reaction. A recurring tale that popped up in the literature was that of an ancient Egyptian pharaoh by the name of Menes, who died following an extreme allergic reaction to a wasp sting in 2640 BC. On first glance this seemed entirely plausible, but a bit of evidence-excavation uncovered an apparently apocryphal tale. According to a 2004 article in the reputable journal *Allergy*, the story is based on one historian's 'fictional account' of Menes' death and has never been accepted by Egyptology scholars. Yet despite being robustly rejected within its own field, this little nugget of historical fiction has taken root in the medical literature on allergy. In truth, Menes may never have existed. And whether he was man or myth, there are two other alternative accounts of his demise: one involving death by hippopotamus, and the other a convoluted tale of being killed by dogs or crocodiles.* However, no wasp anywhere.

For that unique, allergy-blaming account we need to meet Lieutenant-Colonel Professor L. A. Waddell, Professor of Chemistry and Pathology, interpreter of hieroglyphics and author of 'virulently racist writings'. On reviewing a pair of partially preserved wooded plates from an alleged Menes burial site, Waddell interpreted a half-circle hieroglyph with a 'sting' as being a wasp. However three other 'wasps' on the plates had no sting and subsequent scholars have agreed the 'wasps' were actually Egyptian numbers. Thus it appears Waddell's hieroglyphics reading level was only mildly more advanced than his understanding of race relations. Indeed a review of his translation skills in other ancient languages described him as having 'immense delusions' and committing 'unpardonable mistakes'. Which may explain why his work is now largely only beloved by 'white supremacists, esoteric scholars and conspiracy theorists'.

In all likelihood the supposed first written account of allergy in human history is in fact just a scratch in a panel of wood interpreted in a rather odd way by a rather odd man. This seems to be supported by the reflections of Waddell's own obituarist, who described volumes of his work as

---

* With a variation on the theme involving being rescued from the dogs by a crocodile. Dubious.

'conjectures ignoring all principles and results of sober research'.* Hoping to learn from history rather than repeat it, I must confess that despite much sober research I couldn't find a reliable source for the first written mention of allergy. So rather than follow in the footsteps of Waddell and relay a fantastic but fictional tale we'll just have to settle for saying the first allergic reaction occurred A Long Time Ago.

IgE and allergy

Today when we think of allergies we usually think of hay fever, pets and peanuts, causing snotty noses and itchy eyes. All of these allergies are due to type I hypersensitivity reactions, which are fuelled by the antibody IgE. A classic example is an allergy to pollen, which is known as hay fever.[†] It all begins when a susceptible person first encounters the pollen and rather than seeing it as the harmless precursor to plant sperm that it is, their immune system identifies it as a threat. We don't know why this happens, but in these unfortunate individuals B cells are stimulated to make large quantities of IgE targeting the pollen.

The first time they encounter the pollen there may be no signs of the burgeoning allergy because, as we discussed earlier, it takes a bit of time for the B cells to craft a specific antibody in large quantities. However the newly minted anti-pollen antibodies are causing mischief even if we can't see evidence of it yet. They travel round the body and latch on to immune cells called mast cells anywhere they find them. This process means the person is now 'sensitised' to the pollen and the primed mast cells lie in wait throughout

---

* All the quotations about Professor Waddell mentioned here come from two sources: the *UCL Institute of Archaeology Bulletin*, http://www.archaeologybulletin.org/articles/10.5334/bha.20106/ and the *Scottish Historical Review*, http://www.jstor.org/stable/25525780?seq=1#page_scan_tab_contents

† Which is odd really, because it's not caused by hay and it doesn't produce a fever.

the body, from the gut to the skin to the lungs, ready to be activated. The next time the person meanders through a pollen-filled field or has a picnic in a gorgeous garden, they enter an instant itchy, snotty hell. This is tripped by the pollen attaching to the IgE-primed mast cells and, like pulling a pin on a grenade, causing them to unleash their allergy-inducing chemicals.

There are two distinct stages to the reaction: the immediate phase and the late phase. The almost instant assault of the immediate phase reaction occurs within minutes as the dirty bomb-like explosion of the mast cell fills the local area with a variety of rapidly acting chemicals. Perhaps the most potent in the mix is histamine, which has a broad range of effects because it can act on a broad range of tissues. It makes small blood vessels dilate and become leaky, causing inflammation and increasing the blood flow to the area. It also works its magic on pruriceptive neurons, a special type of nerve fibre dedicated to making us feel itchy. Thanks for that one, evolution.

The consequence of these effects depends on exactly where the allergen has caused the histamine release. So in the case of pollen it tends to get into your eyes, sinuses, throat and nose, where the inflammation and irritation causes itchy eyes, sneezing and streams of snot. Conversely, people with allergies triggered by things they eat experience the joys of histamine in their digestive system – think abdominal pain, vomiting and diarrhoea.

The late phase reaction rears its inflammatory head several hours later when white blood cells, including neutrophils, are recruited to the area. As with the early phase the impact of this stage in the allergic reaction depends on where it was triggered. So in the skin, the late phase produces swelling, pain, warmth and redness, while in the lungs it causes an increase in mucus production. Unlike the earlier effects, which tend to resolve relatively quickly, the late phase is the reaction that just keeps on giving – it can take one or two days for the process to come to a halt.

Why do we get allergies?

So why are some of us blessed with an immune system that decides innocuous things like pollen are worthy of all-out warfare? Part of the answer lies in genetics. Since the early 1900s we've noticed that allergies seem to run in families, and in 1923 Arthur F. Coca and Robert A. Cooke proposed the term 'atopy' to describe immediate hypersensitivity reactions with an inherited element. While there were flaws in Coca and Cooke's understanding of atopy (they thought a single gene was responsible), the term has stuck and is now often used to mean an inherited tendency to make IgE antibodies to allergens. Atopic individuals are more likely to have hay fever, as well as allergic forms of eczema and asthma. The pattern of inheritance is a complex one, which means we can't look at someone's DNA sequence and say with any certainty that they will develop these diseases. This is because there are lots of different genes involved and their impact on health is affected by the way the different genes interact, not only with each other, but also with the environment the person is in.

The impact of the environment on allergies is not to be sniffed at. In 1989 an epidemiologist by the name of David Strachan was working on a puzzling public health problem. With greater adoption of vaccines and improvements in sanitation and hygiene, modern medicine seemed to be making strides forward in the battle against infectious diseases. However, while rates of many infections were dropping, cases of allergies in Western countries seemed to be on an inexorable rise. This upward trend continues in recent times – peanut allergies in American children increased more than threefold from 0.4 per cent in 1997 to 1.4 per cent in 2010. If we consider food allergies more broadly, we find an 18 per cent increase among under-18s from 1997 to 2007. This increase has led us to a world where an estimated 50 million Americans suffer from some kind of allergy, making it the fifth most common chronic disease in the US.

Why? Well, back in the eighties Strachan proposed that a rise in allergies could be due to a decline in our exposure to dirt and germs. The basis for his theory was a study he conducted of over 17,000 British schoolchildren, which found that children with a greater number of older siblings seemed to have a lower risk of hay fever. Strachan hypothesised that having older siblings exposed the children to various germs; the greater their number of older siblings the more germy their environment became, and this made them less susceptible to hay fever. The idea was branded the hygiene hypothesis and at the time came out to mixed reviews, primarily because there was no known mechanism to explain how this paradox of 'more germs are better' could possibly be true. However, the immunology community became more convinced when it emerged that the subtype of T helper cells that help battle bacteria and viruses (Th 1 cells) can also inhibit the other subtype of T helper cells that seem to promote allergies (Th 2 cells). So perhaps being exposed to microbes helps tip the balance of the immune system see-saw away from a pro-allergic state. The hygiene hypothesis also fits with other pieces of the puzzle, like the finding that children who grow up on farms exposed to a cornucopia of dirt, germs and animals are less likely to have allergies. Moreover, the idea that the environment is key has been supported by studies that have found that immigrant children in the US are more likely to have allergies the longer they have resided in the States. However, while it's compelling, the evidence for the hygiene hypothesis isn't without question marks. For instance, there are outliers in the hygiene–allergy trend, such as Japan, which has the same modern approach to hygiene as the US but has a much lower incidence of adult asthma. So we need to look for additional explanations for the rise of allergies.

Another intriguing allergy phenomenon is best demonstrated by an unpleasant question: what does a fillet steak have in common with a tick from Texas? The answer is that a bite from the tick may leave you with a life-threatening allergy to the steak. The Lone Star Tick is found in many US states, from Texas to Maine, and gets its name from the resemblance of the single white dot on its back to the lone star found on the Texan flag. When this creepy little critter bites someone, it hangs on with teeth like tiny hooks and spits a chemical mix of its own body fluids into the wound to keep itself firmly attached to its new host. An emerging theory among immunologists is that a compound from the tick's gut, called Alpha Gal, induces an immune response from the victim. This all seems pretty reasonable; frankly, developing an immune reaction to tick-gut spit sounds eminently sensible. However Alpha Gal is also found on red meat, and the theory is that the antibodies designed for the tick's Alpha Gal can also attach to red meat. So the next time a person eats a steak (or any other type of red meat) they are effectively sensitised to it and experience an allergic reaction. The symptoms can range from vomiting and diarrhoea, to a life-threatening drop in blood pressure and difficulty breathing. There is no known cure, which means those who develop the allergy are condemned to avoid steak, burgers, bacon and all other red-meat-related fun for ever.

It's worth noting that at present the tick–meat connection is more theory than established fact – scientists have demonstrated Alpha Gal in the gut of some ticks, but not as yet in the Lone Stars. However, the concept of allergic cross-reactivity is well established. It's the phenomenon where different substances contain either the same allergy-triggering molecule, or at least one so structurally similar that it triggers the same response. For example, people allergic to peanuts may also be allergic to other members of the legume family like soya, lentils and beans. Similarly, people who are allergic to pollen can suffer from pollen–food syndrome because proteins in pollen are very similar to proteins in a broad range of fruit and veg. If a susceptible person eats a peach their throat and lips can become itchy and mildly swollen. Thankfully cooking can transform the shape of these proteins, so it tends to be raw fruit and veg that triggers the pollen–food syndrome reaction.

Cross-reactivity is also the reason that people who are allergic to kiwi fruit might want to think twice about condoms being the way to ensure safe sex. This is because latex-containing condoms, kiwi fruit, carrots, apples, bananas and avocados are all connected by a common cross-reacting allergen, so people allergic to one may be allergic to them all. Thankfully latex-free condoms mean people with such allergies can still enjoy allergy-free sex.

Unusual allergies

According to the US Centers for Disease Control, 90 per cent of all allergies can be attributed to just eight things: milk, eggs, peanuts, nuts from trees, fish, shellfish, soy and wheat. While these may be the most important allergens, this isn't a textbook, which means we have the luxury of a little detour through the weirdest ones: exercise, water, high heels, the cold, stale pancake mix and scratching.

UK national gymnast Tasha Coates earned an impressive five medals at the British Disability Gymnastics 2014. What makes this all the more incredible is that she achieved this despite being allergic to exercise. Not in the sense that I use when my husband is encouraging me to take up running, but in an actual 'life-threatening condition that landed her in hospital 30 times in 12 months' way. Tasha's official diagnosis is a disorder of her mast cells that makes them release histamine and other chemicals in response to a variety of triggers, including exercise. This can cause her airways to constrict, leaving her struggling to breathe and at risk of death. Medicine still has much to learn about mast cell disorders like Tasha's, and the actual mechanism underpinning why her mast cells do this in response to exercise is a bit of a mystery. The same is true of another rare condition called food-dependent exercise-induced anaphylaxis (FDEIA). People with this disorder have an allergic reaction when they combine exercise and a triggering food within a few hours of each other. The food varies between sufferers, with case studies in the medical literature covering everything from seeds to snails. The reaction can be fatal, so people with FDEIA are advised either to avoid the offending food altogether, or to avoid exertion four to six hours after eating it.

While being allergic to exercise is definitely a fairly fundamental challenge, there is something even more essential to life some people are allergic to: water. Hang on, you may be shouting, aren't we mainly made up of water? Indeed, newborn babies are thought to be the most watery human beings of all at a whopping 78 per cent water. In water-content terms that puts them somewhere between a banana and a tomato. By the time we're adults, we're only 55–60 per cent water, which is closer to cooked beef. However, the point holds that by any measure we're made up of a lot of water, so how can anyone possibly be allergic to it?

Well, water allergy is a physical urticaria, which means it's the physical contact with water on the skin that causes the raised red rash known as urticaria. The water inside our cells is not a problem, however, when water is in contact with the surface of the skin, irrespective of its temperature or source, it leaves an eruption of lesions in its wake. We don't fully understand why this happens – some researchers think the water is helping another unknown allergen to penetrate the skin, while others suggest water reacts with the upper layers of skin to create an unidentified compound that activates mast cells. The truth is we don't currently know the cause, but we do know the impact it can have on patients, turning something simple like a shower or bath into an itchy, rashy nightmare. Thankfully this disorder is exceedingly rare, with fewer than 100 cases of water allergy, or 'aquagenic urticaria' to give it its proper title, having ever been described. It's diagnosed by applying a water compress to the skin and seeing if the lesions appear. It's important to exclude other unusual allergies, so patients have to undergo a range of slightly medieval tests. For instance a 6kg (13.2lb) weight may be applied to the skin for 20 minutes to see if pressure is the cause, or an ice-cube-filled bag may be held to the forearm for 20 minutes to rule out a cold allergy. Similarly, alcohol or other organic solvents can be applied to the skin to check whether the reaction is specific to water. For those rare cases where aquagenic urticaria is diagnosed there are at least some treatment options available, for instance in some people strong antihistamines keep the symptoms at bay, while others have benefited from UV light therapy. Just as well, because avoiding showers for life would have a whole range of awkward social side-effects.

Speaking of the social impact of unusual allergies: ladies (or indeed gentlemen), imagine being allergic to your handbag, your heels and your mascara. This is a reality for people with a potassium dichromate allergy, because this chemical is used in a wealth of different processes including traditional leather tanning and cosmetics manufacture. Trying to avoid potassium dichromate is actually incredibly difficult because it crops up in everything from cleaning products, to cement, to ballpoint pen ink. The mechanism underlying this allergy is a bit different from the IgE–mast cell system we met at the outset of this chapter. Potassium dichromate allergy is an allergic contact dermatitis, which is a type of skin rash caused by T helper cells acting in a most unhelpful manner. On contact with the potassium dichromate, sensitised T helper cells start to secrete interferon-gamma, which is one of the most potent activators of macrophages. Many macrophages consequently flock to the site and begin spilling their own inflammation-inducing chemicals into the area. The upshot of this inflammatory process is red, inflamed, blistered, dry, thickened and cracked skin.

Potassium dichromate isn't the only allergen that can trigger this type of reaction – approximately 3,000 chemicals have been found to cause allergic contact dermatitis. Just 25 of these chemicals account for about half of all cases of allergic contact dermatitis. The number one culprit worldwide is nickel, which is found in jewellery from the cheaper end of the counter as well as luxury electronic devices. Fitbit, makers of wristbands that measure physical activity, are acutely aware of this fact. In 2014 they began the withdrawal of more than a million of their devices for causing allergic contact dermatitis due to a mix of nickel and an adhesive called methacrylate. Those affected can at least take comfort from the fact that Fitbit is worn on the wrist. Individuals with intimate piercings, on the other hand, can discover the presence of nickel in their jewellery in a most unpleasant way.

Dermatologists are the Sherlock Holmes of the medical world. Sitting on the sofa right now, I'm wearing clothes of several fabrics laundered in a multitude of chemicals, eating a meal with over 20 ingredients and breathing air carrying pollen, dust and pollutants. And I have a cat sitting next to me. When patients come to the dermatologist with a rash it's the dermatologist's job to hunt down the guilty allergen from the many, many things we come into contact with every day. Sometimes it's easy, such as a rash exactly where a new nickel-containing ring sits. Sometimes it's not. Let's have a go with a real case that falls into the latter camp. A nine-year-old otherwise healthy girl comes to the doctor complaining that three days ago, after drinking a strawberry slushy, her lips became swollen and it became hard to swallow. Strawberry allergy perhaps? Well, a few weeks ago she also had a rash all over her body following a shower after swimming. Water maybe? Hmmm, but that doesn't fit with the third piece of the puzzle: her skin becomes red and itchy after walking through the freezer aisle of the local grocery store. Have you worked out the cause? She had an allergy to the cold. Like the water allergy we encountered earlier, this is a physical urticaria. Exposure to the cold triggers mast cells to release histamine, which in turn produces the swelling and rash the little girl described. In her case the symptoms cleared up of their own accord, but for other people the only option is to avoid the cold and have antihistamines to hand. As one practical mum of two children with cold allergies noted, 'they have to be kids', which means they get to make snow angels, but wearing a lot more layers than other kids and with a bottle of Benadryl antihistamine on standby.

Can you write on your skin with nothing more than your finger? People with dermographism can. Dermographism is another example of a physical urticaria. However, rather than water or cold being the trigger, a simple scratch or stroke of moderate pressure leaves the skin with a raised red line where it's been touched. As a Google image search attests (warning: not for the faint-hearted), this can create the appearance of a human Etch A Sketch. Thankfully for many people the symptoms can be controlled with antihistamines and the lines tend to disappear within 30 minutes.[*]

Stale pancake mix definitely sounds unpleasant, but add in an allergy and for one 19-year-old man it proved fatal. He was making breakfast with friends, when they found a box of pancake mix. It was two years out of date, but they figured it would probably be fine and made themselves some pancakes. A few bites in, his friends decided that pancakes probably shouldn't taste like 'rubbing alcohol' and stopped eating them. He decided to continue munching, and made his way through two whole pancakes when he suddenly became short of breath. His friends rushed him to hospital, where he sadly died. Analysis of the pancake mix found four types of mould: *Penicillium*, *Fusarium*, *Mucor* and *Aspergillus*. The victim had a history of mould allergy and on that day he consumed a massive dose of mould, leading to a massive allergic reaction, known as anaphylactic shock.

Anaphylactic shock, also known as anaphylaxis, is a potentially fatal reaction that causes a person's airways to narrow and their blood pressure to plummet. About 25 per cent of people with a food allergy will have a near-fatal anaphylactic reaction in their lifetime. IgE and mast cells are to blame for this severe reaction, which is caused by the mast cells releasing their chemical granules en masse across the entire body, unleashing the allergy equivalent of a weapon of mass destruction. As blood vessels across the body dilate and become leaky, up to 50 per cent of the volume of fluid contained inside those blood vessels rushes out into the surrounding tissue in a matter of minutes. This means there simply isn't enough blood to maintain a normal blood pressure and deliver oxygen and nutrients to the brain at the right rate. The influx of fluid also makes the tissues swell rapidly and when this happens in the throat it can become difficult to breathe. This can be made worse by spasms in the airway which cut off the victim's oxygen supply completely. Knowing all of this is more than morbid curiosity – knowing this has

---

[*] Shaking doesn't help, unlike with an Etch A Sketch.

helped save lives. We can counter the known effects of anaphylaxis by giving adrenaline, a drug which relaxes the airways and blood vessels. People who are known to be at risk of anaphylaxis should carry something like an EpiPen, which is a super-simple way of injecting adrenaline fast. Despite this knowledge, four people in the US will die this week, and every other week, from an anaphylactic reaction. So what else can we do to prevent anaphylaxis and other allergic reactions?

What can we do about allergies?

The most simple answer is to avoid the allergen. However, in practice this is often distinctly complicated. Take pollen – *it's floating in the air*. Short of donning an industrial-strength respirator before leaving the house every day, it's simply not possible for people with hay fever to avoid inhaling the source of their irritation. That said, they can take steps to lessen their exposure by avoiding going outside when pollen levels are particularly high. This is made possible by the pollen count – an announcement that usually accompanies the weather forecast on the TV and on the internet throughout the summer months when pollen levels peak. Calculating the pollen count is a highly technical affair. In the UK, the BBC's pollen trap is essentially a piece of sticky tape inside a box with a slit in it, placed on the roof of a building. Pollen gets stuck to the sticky tape, which is then put under a microscope so someone can literally count the pollen. Despite our being able to put a man on the moon, apparently nobody has yet developed a satisfactory method for automating the pollen count.

That said, the very existence of the pollen count is a remarkable thing – an allergy warning that gets its own TV slot is a unique and exceptional achievement. Fittingly, the man responsible is just as unique and exceptional: William Frankland, granddaddy of all things allergy. Born the same year as the *Titanic* sank, Frankland has been a doctor to Saddam Hussein (who thought he had a mould allergy and asthma, but didn't), worked under Sir Alexander Fleming (famed for discovering penicillin), and at the age of 99 was the oldest ever expert witness in a court case. While working for Fleming he correctly predicted that increased use of penicillin would see a rise in penicillin allergies. However it wasn't just his brilliant mind, but his dedication to his craft that made Frankland the foremost expert in his field. He had a penchant for self-experimentation and in the 1950s allowed a South American insect by the name of *Rhodnius prolixus* to bite him and suck his blood once a week. By the eighth week he had a full-blown anaphylactic reaction and had to be injected with adrenaline. Two hours later he was back to seeing patients, but then had a further delayed reaction requiring another injection of adrenaline before driving himself home at the end of the day's work. Today he is just as ebullient, and as a centenarian is still attending allergy conferences, publishing academic research papers and delivering lectures to inspire the next generation of immunologists.

Avoiding pollen is one thing, but what about being allergic to something you really want to touch? A few years ago I decided to get a cat. I really wanted a dog but that was impractical so I looked up the biggest breed of cat it's legal to purchase in the UK. That turned out to be a Savannah cat, which is something akin to a small leopard, and looked like it might eat the neighbour's Jack Russell as an amuse-bouche before moving on to their toddler for the main event. So instead I dragged my husband to meet some Maine Coons – picture 7–11kg (15–25lb) of fluffy, purring gorgeousness. Sadly for my husband it was 7–11kg of itch-inducing, eye-watering misery. He was allergic to them. Sad face. However, as the old adage goes: when life gives you lemons, use it as an excuse to see a lot of kittens. I made a list of cat breeds in order of descending size and we set about meeting them to see if there were any hubbie wasn't allergic to. Which is how we came to own Loki the Ragdoll and Serafina the RagaMuffin – members of the fourth and fifth largest cat breeds in the UK.

Other people* take a different approach to combining pets and allergies. Some of these people find themselves on a quest

---

* Possibly with less tolerant husbands …

for a bald cat. The logic goes like this: I'm allergic to cat hair, so all I need is a cat without the hair. Such curiosities exist under the suitably rarefied name the Sphynx. The Sphynx breed of cat originated in the 1960s when a furry feline gave birth to a kitten with a mutation that left it hairless. Since then the breed has developed and Sphynx kittens come in a range of states of hairlessness. Many have a little hair around their ears, while others are covered in a very fine down that reportedly makes them feel like a warm peach. Their wrinkly skin has been compared to butter, a warm chamois or a suede hot-water bottle. However, if you have a cat allergy and want a buttery peach of your very own, be warned: even a totally naked Sphynx is no guarantee of an allergen-free kitty-cat. This is because it's not hair that's the key culprit in cat allergies. Instead there's a protein called Fel d1 which is found in cat fur, saliva, urine and skin glands. While the absence of fur reduces the allergen levels, contact with Fel d1 is still inevitable. It's possible, indeed encouraged, that owners give their Sphynx a bath to reduce the amount of allergen on the cat's skin but some people will still suffer allergic reactions to even very low levels of Fel d1. This creates a marketable niche for mutant cats. One American company has created a cat that produces a mutated form of Fel d1 which they claim is less likely to cause allergies. However prospective owners might find they have an allergic reaction to the price: almost $9,000 (£6,800) for a standard size cat or $30,000 (£22,700) for a large cat.

So, a hairless cat is no guarantee of an allergy-free life, a mutant cat might break the bank and a skinless cat is neither biologically feasible nor child-appropriate. It's enough to drive a cat person to become a dog person. Sadly dog allergies are also a common and troublesome problem. Some people just decide not to have a pet at all, but what if a pet is a need rather than a want? This was the dilemma of an American woman who was visually impaired and in need of a guide dog, but her husband had a dog allergy. She contacted Wally Conron, breeding manager at the Royal Guide Dog Association of Australia, who came up with a creative solution: the Labradoodle. The Labradoodle was a new cross-breed, blending the guide-dog skills of the Labrador with the minimal fur-shedding of a standard poodle. In many ways the breed was a success, becoming hugely popular around the globe and boasting celebrity owners like Jennifer Aniston and Tiger Woods. So why does Wally Conron wish he'd never invented them? Well, they have been credited with the rise of the designer dog. Now you can get a range of Poodle crosses: Cockapoos, Yorkipoos, I'll leave the name of the Shih-Tzu cross to your imagination. Conron believes this has led to unhealthy and abandoned animals bred for profit (a Labradoodle puppy can cost as much as £1,500 ($,1971)). Moreover, the challenge of a cross-breed is the unpredictability of the puppies created – they might be a bit more 'labra' or a bit more 'doodle', and that affects the kind of coat they have and how much it sheds. Sadly the original guide-dog breeding programme was eventually abandoned because of the inconsistent results. Stick insect, anyone? Actually, there's some evidence people can develop a sensitivity to stick insects too. Sea monkeys, anyone?

Nothing demonstrates the challenge of trying to avoid an allergen better than the phrase 'may contain nuts'. Have a look at some jars in your kitchen – how many have that on the label? Or perhaps a variation on the theme 'made in a factory where nuts are used'? Now imagine the life of a peanut allergy sufferer. You might think the packaging is just covering itself by trying to pre-empt the remotest possibility of a lawsuit, but the truth is that people with peanut allergies can be exquisitely allergic. Like four-year-old Fae. Fae stopped breathing on a flight home from her summer holidays because of a packet of peanuts belonging to another passenger. He didn't feed them to Fae, he didn't even hand them to her. He was sitting four rows behind her and simply opened the packet. Fae's mum had warned the cabin crew about her daughter's allergy, and they in turn had stopped the sale of nuts on the flight and put out three warnings to passengers advising them not to eat their own. Perhaps owing to a language barrier, perhaps because of an underestimation of the danger, a male passenger ignored these warnings and it seems the plane's air-conditioning system circulated the peanut particles. Thankfully Fae had an adrenaline injection with her and a passenger who was also an ambulance driver used it to save her life. The male passenger was banned from flying with Ryanair for two years.

Not all airlines are so proactive in ensuring the safety of their peanut allergic customers. This has led a determined American mum to launch a petition for a 'bill of rights' for food allergic customers, which would include creating allergen-free buffer zones around affected passengers. However, for now, most airlines adopt the same approach as food manufacturers, and peanut allergy sufferers are faced with flights that carry the label 'may contain nuts'.

From pollen to peanuts, it seems that just avoiding allergens is not that easy. Even billionaire Sean Parker, founder of Napster and Facebook's first president, has been unable to craft a world where he avoids peanuts and shellfish, with his allergies repeatedly landing him in the emergency room. However, unlike the rest of us he can give $24 million (£18 million) to Stanford University to fund the Sean N. Parker Center for Allergy Research. The aim of the centre is not only to develop novel treatments, but to deepen and extend our understanding of the immunology of allergies with a view to creating a cure.

Parker isn't the only one to 'like' the idea of a cure, and some scientists are actually making exciting progress towards this goal. In 2014, a team of researchers at Cambridge University published the results of a world first. In their clinical trial, 85 peanut allergic children aged 7–16 years old were allocated to either the traditional 'peanut avoidance' approach or a new treatment called oral immunotherapy (OIT). This wordy experimental treatment has a simple premise: can exposure to increasing doses of peanut protein desensitise allergy sufferers so they stop reacting to peanuts? What they found was yes, with a couple of caveats. After six months of eating very controlled amounts of peanut protein,

62 per cent of the children on the new treatment could eat the equivalent of 10 peanuts without reacting and at least 84 per cent could eat five peanuts without ill effects. This compared to zero per cent of the children who had continued to just avoid peanuts. Five to ten peanuts is above the level found in contaminated snacks and meals, so this is a fantastic result which could point the way to a world where peanut allergy sufferers (and their parents) don't live in constant fear of accidental ingestion and life-threatening reactions.

The trial represents an exciting step forward but, as mentioned, there are a couple of caveats. First, it's still a relatively small number of people. Before we get too excited it would be good to see the same result in a trial with more people. Of course there are significant risks involved in running these kinds of trials. While nobody in the Cambridge study suffered a serious adverse reaction, one child was withdrawn from the study after they needed to inject themselves with adrenaline when they began wheezing after taking their OIT. It must be remembered there is a real risk of anaphylaxis and death associated with exposing these people to peanuts, so these studies can only be carried out in specialist centres with very controlled conditions. So even if it is a success in a larger number of people, because of the level of control required, OIT isn't the easiest of treatments to administer. The second caveat is that other evidence suggests that if you stop the peanut consumption after a few months, the effects don't last. The study team think it could be several years before all the B cells in the bone marrow that once made IgE to attack peanuts die off. Ultimately nobody knows how long patients would need to keep consuming peanut protein to ensure continued protection against allergic reactions from accidental peanut exposure. A study of sufficient length hasn't been done. Yet.

All of the allergies we've discussed involve an overreaction to something harmless, be it pets, pollen or peanuts. While the body has failed to spot that the thing is harmless, it is at least correct in recognising something foreign. When it fails to make that distinction correctly – the one between self and foreign – we're in real trouble. Then the world's most powerful biological defence system can become violently overactive, gnawing away at organs and dissolving bones, destroying the very thing it exists to protect: us. In our next chapter we will unveil the darker side of the immune system, exploring the potentially deadly phenomenon of autoimmunity.

CHAPTER TWELVE

Auto-immunity: The Science of Friendly Fire

What secret powers of healing might the humble human foreskin be hiding? If the Terminator had a pancreas, what might it look like? And why does a cut of meat called the mutton flap threaten the island nation of Tonga? These bizarre and seemingly disparate questions are all bound together by a common disease of international importance: diabetes. Diabetes offers a cautionary tale of what can happen when our body's defences turn on us. Through its example we'll expose the darker side of the immune system and see how medical advances might stop the terrible damage caused by less-than-friendly fire. Our story of immunological misadventure and masochism begins in a land far, far away, with the fattest king in recorded history ...

The late King Tupou IV of Tonga holds the Guinness World Record for having been the planet's heaviest monarch. At his most royally rotund he tipped the scales at 209.5kg (462lb). If His Majesty sat on a see-saw, balancing it would have necessitated loading the other side with either a) a particularly hefty male specimen of that king of cats, the African Lion or b) perching roughly three and a third copies of British monarch Queen Elizabeth II opposite him.

Blue blood or not, being that overweight isn't good for your heart. No doubt aware of the risks to his health, the king succeeded in losing about 70kg (154lb) through 'eating a lot of soup'. The king could have sculpted a fully grown man out of the fat he lost from his body, which is an impressive, if mildly perturbing, feat of weight loss. However his work was far from done because not only was King Tupou IV the most obese monarch in recorded history, he also presided over one of the most obese nations on the planet. Alas, his attempts in the 1990s to lead his people on a similar journey of weight loss were about as successful as his decision to put a court jester in charge of investments.[*] In 2014, eight years after King Tupou IV's death, a study in the *Lancet* estimated that 83.5 per cent of Tongan men and 88.3 per cent of Tongan women over the age of 20 were overweight. These statistics earned Tonga the unenviable title of most obese country in the world.

There were various explanations for this sizeable problem, including a culture that venerated the larger figure as synonymous with wealth and beauty. However, one of the biggest problems for the national waistline has been the evolution of the national diet. As you might expect for a tropical island surrounded by a bountiful blue sea, the traditional Tongan diet is rich in fish and root vegetables. It's the kind of thing you might eat in January, that month of guilt when you try and compensate for the over-indulgences

---

[*] In fairness to the king, his court jester Jesse Bogdonoff was also a former American banker. Sadly their joker lost the country an estimated \$26 million (£21 million) in failed investments.

of the December holiday season by Being Good. However this nutritional goodness has fallen from favour in modern-day Tonga and been replaced with the rise of a new staple: the appetisingly named 'mutton flap'. Mutton flaps are the dietary work of the devil. This cheap cut of meat comes from the end of the sheep's ribs and is the sort of food that would make any self-respecting dietitian drop to the floor in the foetal position clutching a bunch of carrot sticks.

According to a Tongan health planning officer interviewed by the BBC, it wouldn't be unusual for a Tongan to enjoy a hearty kilogram (2.2lb) of mutton flaps in a single sitting. An incredible 40g in every 100g of mutton flaps consists purely of fat – that's four times more fat per gram than a Big Mac – and means a single serving of mutton flaps contains more fat than the average adult should eat in five days. In fact the health planning officer himself professed to having eaten such portions before going on a diet. One of the triggers for his weight loss was getting a diagnosis of diabetes, a disease that renders the body unable to control blood sugar levels. Sadly, this is not an unusual experience if you're a Pacific Islander, with a government report estimating that diabetes affects up to 40 per cent of the population of the island nations.

Of course diabetes is not limited to the Pacific Islands or consumers of mutton flaps; in fact it's one of the most common diseases on the planet and a major threat to human health. More than four million people in the UK are currently living with diabetes, with one person being newly diagnosed every two minutes. This costs the British National Health Service an estimated £1.5 million ($1.98 million) an hour, which over the course of a year means diabetes consumes 8 per cent of the entire NHS budget. The problem of diabetes is even bigger in the US, where 9.3 per cent of the population has the disease, rising to over 25 per cent of the over-65s. Diabetes is not just a common disease, it is also a deadly one. It represents the seventh leading cause of death in the States. It kills more Americans than tornadoes, terrorism and shark attacks combined. It kills even more people in the US than guns do (excluding suicides). The global importance of understanding and treating diabetes is underscored by the fact that by 2040, an estimated 642 million people worldwide will have diabetes. That's one in ten people on the planet.

Yet despite diabetes being such a formidable and increasingly common problem, the scientific community is still scratching its collective head over the question of how this sickly sweet disease arises. One of the prime suspects is the immune system. Unlike most of the cheerleading chapters in this book, here the immune system is cast as the villain of the piece for performing an act of self-harm known as autoimmunity. Autoimmune diseases reveal the dark side of the immune system, with T cells, B cells or antibodies turning on the body they were designed to protect and tearing up perfectly healthy tissue. There are over 80 different known autoimmune diseases attacking every part of the body from knees to nerves, glands to gonads.

Type I diabetes

There are two main forms of diabetes, Type I and Type II. In the case of Type I diabetes there has long been damning evidence that the disease results from the immune system launching an attack on the pancreas. The pancreas is a floppy, feather-shaped organ that nestles behind your stomach and acts as the commander-in-chief when it comes to controlling blood sugar levels. Essential to this control are the beta cells of the pancreas, which secrete a hormone called insulin. In a healthy pancreas insulin is released into the blood in controlled amounts so that irrespective of whether you haven't eaten for hours or have just consumed enough sugar to rot the collective grin of a boy band, your blood sugar stays within the normal range. However, the damage to the pancreas in Type I diabetes means the beta cells stop producing insulin, thereby allowing blood sugar levels to soar to harmful heights. In the short term the body tries to offload the excess sugar in one of the only ways it knows how: pee. This leaves diabetics needing to pee a lot (one of the most common symptoms of diabetes), and explains the disease's full Latin name: *diabetes* 'to pass through' *mellitus* 'honey sweet'. However in the long term having high sugar levels damages small blood vessels throughout the body, which can cause a litany of serious problems from blindness to gangrene. Equally, when it's been too long since someone with diabetes last ate, their blood sugar levels plummet. Their body is left trying to run on empty, making them tired, shaky and sweaty. Unchecked, low blood glucose can lead to unconsciousness and ultimately death.

While insulin harvested from pigs was once the height of our diabetes-defeating prowess, today patients are enrolling in trials of a bionic pancreas that lets them track their blood sugar levels via their Apple watch. One of the engineers leading on the bionic pancreas is Professor Ed Damiano, a biomedical engineering expert from Boston University who was driven to develop the device by his own son's experience of diabetes. Working with doctors from Massachusetts General Hospital, Prof. Damiano hopes to have the device ready for commercial release by the time his now 16-year-old son heads off to college. The Terminator-worthy bionic pancreas is called iLet[*] and it analyses the patient's blood sugar level every five minutes. In response to the reading it releases a precise amount of either synthetic insulin, to bring the blood sugar down, or another hormone called glucagon, to push it up. The bionic pancreas will have to prove itself in clinical trials before it becomes a mainstream treatment, but it will also have to take on a threat that traditional treatments don't: hackers. At present cybersecurity for medical devices is largely reserved for handy plot twists in TV shows like *Homeland*, where the vice-president of the US was killed when his pacemaker was hacked. However in 2015 the FDA strongly encouraged hospitals to stop using a computerised drug delivery pump because of cybersecurity vulnerabilities that would allow someone to hack the pump's software and

---

[*] In homage to the islets of Langerhans, the bits of the pancreas that include the insulin-making beta cells.

change the amount of drug it delivered. If it were possible to hack the bionic pancreas, the results could be deadly.

Impressive as iLet is, even more desirable would be a healthy pancreas that doesn't need a firewall. Scientists from the University of California are tinkering with human foreskin cells to provide just that. Using a combination of chemicals, growth factors and genetic reprogramming they managed to transform human foreskin cells into something resembling pancreatic cells, crucially complete with the ability to make insulin. The research team then took mice that had been treated to create a manmade diabetes and grafted the pretend pancreas cells into one of their kidneys. They found that the addition of the tampered foreskin cells stopped the mice having diabetes and when they removed the protective penis-pancreas-kidney cells, the mice duly developed diabetes. Clearly it's a long journey from this small study in mice to a lab-grown pancreas for humans, but it could be the first step down an exceedingly interesting avenue.

The deeper we delve into the mechanisms of autoimmune destruction that underpin Type I diabetes, the smarter our treatment of the disease becomes and the closer we get to a cure. Recent studies have identified five key antibodies in the blood of Type I diabetics that attack various components of the insulin-producing beta cells. The fifth and final beta cell target was identified in 2016, and the hope is that now we have a complete picture of what the immune system is attacking, researchers will move towards preventing the damage, and therefore the disease, from ever arising.

Despite such advances in understanding the immunology of diabetes, the answer to a fundamental question still eludes us: why does the immune system turn on the body it has defended since birth? We know that in a normal, healthy person the immune system is taught a key skill we should all learn very early in life: tolerance. Tolerance is a concept we encountered in our exploration of transplants, and involves teaching the patrolling forces of the immune system to stand down if the cell they're interrogating is a healthy cell that belongs to the body. It's a bit like a border patrol force wandering through the body and checking passports. If a cell shows the correct passport, the immune cells move on; if they don't, the cell is in trouble. Essential to this premise is that the border patrol cells only pick on infected or foreign cells. Unfortunately, the way our multitude of immune cells are made means that some see a healthy passport and attack instead of moving on. The process of tolerance is like a training camp for the newly created border guards, showing them a range of perfectly normal passports representing all of the healthy cell types in the body and ordering those that react incorrectly to self-destruct. This process isn't perfect, so some escape into the bloodstream before they've been eliminated.

There's a back-up plan to pick up these rogue escapee cells in the shape of T regulatory cells. These cells answer the question *Quis custodiet ipsos custodes?* – who guards the guards? The T regulatory cells identify and suppress immune cells with an unhealthy interest in normal cells. While it's still unclear what has gone awry in diabetics to allow the immune system to turn on the pancreas, scientists are trying to kick these regulatory T cells into action to slow the progression of the disease. Researchers on the MultiPepT1De study based in London have trialled a Type I diabetes treatment that exposes the body to fragments of proteins from beta cells with the hope of reminding T regulatory cells to protect any remaining beta cells. Type I diabetics typically have 15–40 per cent of their beta cells still alive and functioning when they are first diagnosed, so the hope is that these cells and their sugar-controlling powers could be saved. The study is still in its infancy, but if it is safe and shows promise of being effective, the drug will be trialled in children to see if it can prevent diabetes ever developing.

Type II diabetes

In some ways Type I diabetes is the minor part of the diabetic tale, as its counterpart Type II diabetes causes 90 per cent of diabetes in adults. While Type I diabetics cannot make insulin, Type II diabetics either have low levels of insulin or their body doesn't respond to it any more. Unlike the clear-cut autoimmune nature of Type I diabetes, Type II is a messy affair with diet, genetics and the immune system all implicated. Of these three, the evidence against obesity is particularly damning. It seems to play a key role in setting up the conditions for diabetes to take hold, possibly by causing chronic, low-grade inflammation that makes the immune system more likely to attack the pancreas. This may explain why the mutton flap diet has been so disastrous for the diabetes rate in Tonga, and why the rise of obesity is propelling us into a future where one in ten of us has diabetes. The role of obesity in Type II diabetes is such that in May 2016, 45 of the world's foremost diabetes experts and organisations came together and issued a statement recommending that diabetics should be offered gastric surgery for weight loss as a standard treatment. As they noted in their statement, this represents a seismic shift in diabetes care rivalled in scale only by the point in history when insulin became available as a treatment.

Our catastrophic diets have also engineered a shift in who gets Type II diabetes. Obesity used to be a predicament of adulthood, and so, once upon a time, was Type II diabetes. Historically it was called Adult Onset diabetes because it was so rarely seen in childhood – prior to 2001 only 3 per cent of adolescent diabetes diagnoses were Type II, with Type I being far more common. However a decade later, riding on a massive wave of overweight and obese children, that figure skyrocketed to 45 per cent and consigned the notion of 'Adult Onset' diabetes to the past.

Of course we are not just what we eat, we are also a mishmash of inherited predispositions and physiological quirks courtesy of our genes. Scientists trying to tease out the role of genetics in Type II diabetes are using an array of techniques, from the very old to the very new. Some use cutting-edge technology to speed-read the genomes of thousands of diabetics and look for common patterns. Others exploit a much older experimental set-up gifted to science by nature: twins. For instance, if you study Fred and George, a pair of identical twins raised together, you know they have the same genetic code *and* had as close to the same childhood environment as two people can have. They went to the same school, lived in the same house, were fed the same sort of things, had the same number of siblings and had parents with the same income to spend on them. Moreover they did all of this at the same point in time so they also lived through the same changes in society more broadly, whether that be living through a new public health initiative or a wizarding war. In Type II diabetes, we know that if one twin has the disease there is a massive 70 per cent chance their identical counterpart will develop it in their lifetime.

Given that identical twins share their genes *and* their childhood environment, you might reasonably ask how we know what fraction of that 70 per cent should be attributed to the relative roles of nature versus nurture. Cue more twins to help us with this problem, this time non-identical twins. Non-identical twins that grow up together have virtually the same environment, but while identical twins share 100 per cent of their genes, non-identical twins only share about 50 per cent of their genes. If one non-identical twin has Type II diabetes there is about a 20–30 per cent risk the other will develop it, which is much smaller than the 70 per cent risk seen in identical twins. Thus genes do seem to have a big impact on Type II diabetes risk, and an avalanche of academic endeavour is currently underway to answer the elusive question of which genes are culpable and how.

The combination of these dietary and genetic risk factors interacting is thought to create a pro-diabetes melting-pot that sets up a state of chronic inflammation in the body. In some people this damages the production of insulin and how the body responds to it, leading to full-blown Type II diabetes. Despite being an exceedingly common disease, the book of Type II diabetes is still being penned, with some fairly basic questions left answered. For instance, while the immune response seems to be involved the exact nature of that involvement is far from clear. Recently some diabetes experts have suggested that, as with the beta cells in Type I diabetes, there could be a specific target that the immune system is attacking. In support of this idea, scientists working on animal studies have identified antibodies which, when transferred from one animal into another, cause the recipient to develop diabetes. However concrete evidence of such a target in people remains elusive and until such a target is found, Type II diabetes can't be classed as a true autoimmune disease.

How the immune system and the environment interact

The complex interplay between the immune system and environmental factors leading to disease isn't limited to Type II diabetes. There are a range of different things that can tip our tiny but deadly defenders against us. It's a bit like having a paddling pool infested with miniature sharks and dipping your toe in the pool: ordinarily the miniature sharks have no interest in eating toes, and just mill about, perhaps conversing about the weather and gobbling the occasional miniature fish. However, if you've cut your toe and blood gets into the water, all bets are off. The blood-tinged water sets the sharks on a feeding frenzy, where fingers, toes and all other appendages are fair game even if they wouldn't normally be on the shark's dinner menu. There are many environmental factors that can act as the blood in our analogy, from obesity, as in Type II diabetes, to perhaps even sunlight levels.

Multiple sclerosis (MS) is a neurological condition caused by the immune system attacking a protein called myelin that sheathes nerves in the brain and/or spinal cord. When the myelin is damaged, the transmission of nerve impulses is disrupted, causing a constellation of symptoms from numbness to tingling, bowel troubles to vision problems. While we don't know exactly why the immune system turns on myelin, studies have implicated a range of risk factors including sunlight and Vikings. Both of these factors relate to the fact MS is more common in very northerly areas; for instance the highest incidence of MS in the world is in Orkney. It has been suggested this may be because of low vitamin D due to limited sun exposure so far north. Another theory is that it could be due to the genetic heritage of people from these areas, hence the Vikings.*

Manmade factors can also tip the immune system in unexpected directions, as Europe discovered when an outbreak of narcolepsy followed a vaccination programme against the 2009 pandemic flu. Over 1,300 people developed vaccine-related narcolepsy, which causes distressing symptoms like falling asleep unexpectedly, sleep paralysis (the inability to move when dropping off or waking up) and cataplexy (loss of muscle control, typically following strong emotions like anger). In 2015, a team of researchers finally found a possible explanation for this unexpected side-effect, as they discovered that the vaccine triggered the creation of antibodies that coincidentally bound to sleep centres in the brain. They concluded that in people who had a genetic predisposition to narcolepsy the creation of these antibodies led to the sleep disorder. To keep this in perspective, over 30 million Europeans were vaccinated, making it an incredibly rare side-effect, affecting only four in every 10,000 of those vaccinated.

Perhaps unsurprisingly, given that a vaccine can trigger an autoimmune response, so can a genuine infection. An example of this was seen in 2016 when Zika virus swept through South America leaving a trail of neurological disaster in its wake. People developed progressively worsening muscle weakness – for most it affected their arms and legs but in extreme cases the weakness spread to the muscles responsible for swallowing and breathing, leaving the person trapped in

---

* While the Vikings suggestion may sound more like click-bait than solid science, our ancestors have more to do with our immune system than you might think. In 2016 a group of researchers concluded that ancient human dalliances with Neanderthals may have left some modern-day humans with an itchy immunological legacy: three Neanderthal genes often found in non-African people that assist in the innate immune response but which also seem to make people more prone to asthma and allergies.

their body and dependent on a ventilator to keep them alive. The culprit was identified as Guillain-Barré, a rare and serious syndrome that is often associated with a recent innocuous viral infection like the common cold, or in this case, the Zika virus.* How a simple infection might turn the immune system against the body and produce this potentially life-threatening autoimmune condition is, to quote our regular refrain in autoimmunity, poorly understood. However the treatment also lies in the realms of immunology as patients are given antibodies from healthy blood donors, which destroy the ones attacking the nervous system.

Throughout this chapter we've discussed diseases caused by overactive immune systems. Yet this story of autoimmunity is just one side of a very dangerous coin, and it's time to flip the coin and explore immunodeficiency – life with a weakened immune system. We will meet people forced to take drugs, try experimental treatments and even live in a bubble to achieve what you and I manage while lounging on the sofa watching Netflix.

---

* Zika is most famous for another reason. While the average person with Zika will experience a mild illness, typified by symptoms including joint pains and fever, the consequences for pregnant women can be horrific. In 2016, Zika hit the headlines for causing babies to be born with microcephaly, which means they have a small head, something that can be associated with abnormal brain development. Consequently pregnant women were told to avoid travelling to countries where there was a high risk of catching Zika.

CHAPTER THIRTEEN

Defenceless: The Boy Who Lived in a Bubble

David Vetter was a much-anticipated and much-loved baby. He came into the world crying and screaming like any other newborn, and for 20 whole seconds he lived in it like the rest of us. Sadly, that's where the normal bit of the story ends. David was carefully enclosed in a specially built plastic bubble where he would spend the rest of his days. The exceptional nature of his existence would lead to incredible scientific discoveries during his 12 years on Earth, but also long after his death. However, perhaps the most important facts to remember are that David's favourite colour was purple, he adored *Star Wars* and he used to hide pencils in an attempt to play hooky from school. He was an ordinary kid thrown into an extraordinary way of life because he had no immune system.

David was born with Severe Combined Immunodeficiency (SCID, pronounced 'skid'). This genetic condition meant David was unable to make his own T cells or Natural Killer cells, and the B cells he made were completely non-functional. He was therefore 'immunodeficient' and unable to fend off the most minor of infections, transforming even the common cold into a potential killer. David's parents were all too aware of the dangers of SCID, having already lost another son, also by the name of David, to SCID when he was just seven months old. After the death of their first son, doctors warned the Vetters that if they conceived another boy there would be a 50 per cent chance he would suffer from SCID. On finding out their next child was a boy the devoutly Catholic couple declined the option of a termination and instead did everything humanly possible to ensure a germ-free birth for their baby. They placed their faith in the ideas of an Experimental Biology professor by the name of Raphael Wilson. He suggested that if a child with SCID could be preserved in a germ-free environment long enough for a matching bone marrow donor to be found, it might be possible to cure them.

Preparations for the birth were almost as involved as trying to plan a military campaign on the moon. Three connecting operating rooms were closed for three consecutive days of intense deep-cleaning. The afternoon before the procedure, signs were erected to divert the normal surge of human traffic away from the corridors nearest the theatres, to minimise the number of germs hitchhiking to the area. The members of staff carefully hand-picked to perform the caesarean section were swabbed and sampled to within an inch of their lives. Any found to be carrying infection were told to stand down and were replaced with a contagion-free colleague. The super-clean team had to work together as one well-oiled machine, with the ability to operate in complete silence while barely moving. This part of the plan was to minimise the movement of air in the room to reduce the risk of microbes parachuting from the air to land on David. The bizarre nature of this requirement shouldn't be underestimated. Operating rooms are normally alive with the sound of communication as surgeons, nurses and anaesthetists exchange instruments, information and ideas. In this case, the team were forced to replace this rich rhythm of conversation with a blanket of silence punctuated by a complex mime of nods and eye movements devised purely for this operation.

The medical team weren't the only ones entering into the unknown. The prospect of giving birth is often stressful, but surely never more so than for anxious mum Carol Vetter in the days leading up to 21 September 1971. She was not exempt from the germ eradication precautions, and had to sleep on sterile sheets, eat sterile food and scrub her entire body with an anti-bacterial soap. Unlike any other expecting mum, surrounded by excited friends and family, she was isolated and only allowed to speak to people on the telephone. When the predefined morning of David's birth arrived, Carol was wrapped in sterile sheets (a feeling she later likened to being mummified) and her tummy was scrubbed for a solid 10 minutes to wipe out any microbes and make it as pristine as skin could ever be.

Finally, preparations for the germ-free sanctum were complete. There was one more unusual step before scalpel could touch skin: 15 minutes of complete stillness and silence. This was to ensure the air would be as motionless as possible to minimise the number of microbes that might be carried on the breeze of a whisper or the gesture of a hand. The time had come. The operation began and, moving slowly but precisely, the surgeons cut David out of his blissfully clean cocoon and hauled his vulnerable little body into the real world.

Newborns are normally swiftly handed to their mum to begin the bonding process through skin-to-skin contact, but David was delivered into the waiting arms of his bubble. The bubble had been prepared as carefully as the operating theatre, with filtered air flowing past a pile of sterile diapers, sterile medical equipment and even sterile holy water. A blood test was taken and for two long weeks the Vetter family waited and prayed, hoping that this David was in the safe 50 per cent. Unfortunately it was not to be and the test results revealed that David had indeed inherited SCID. Devastating as this was for them at the time, the extreme precautions taken by Carol and the medical team had at least given David a chance at life.

Nobody predicted just how unique a life it would be, because nobody anticipated David would wait so long for a matching donor. The team had always seen the bubble as a short-term stopgap, but without a bone marrow match they were faced with an almighty list of unanticipated everyday problems to solve. For instance, how do you make sterile toothpaste? They tried autoclaving it (heating it) to kill any microbes in it, but it turns out super-heated toothpaste has more in common with concrete than a dental hygiene product ever should. They considered bubbling gases through the toothpaste but concluded the most likely outcome was one giant exploding toothpaste bubble. So they didn't do that. The ultimate solution came not from cutting-edge technology, but from the past: salt and baking soda. In the years before commercial toothpaste, this combination was used to keep our pearly whites in tip-top condition and it worked for David too.

David's teeth were also aided by the fact that without an immune system this little boy lived his life having to be shielded from ice cream, cake and Coca-Cola. Up until the age of five his mum would bake a cake and blow out the candles, but the birthday boy himself couldn't eat the unsterilised sugary goodness because there was a very real risk it would kill him. On his fifth birthday a cake was found which could be safely sterilised, and David enjoyed his first mouthful of birthday cake (albeit a vacuum-packed version in a can). Another of David's long-standing ambitions was to drink Coca-Cola. Having watched people on TV enjoying the fizzy drink he asked if he could have some. Staff tried experiments to make his wish come true but the heating process ruined the bubbles and the taste. When David was finally taken from his bubble in the last few days of his life one of his first requests was to finally drink Coca-Cola. Sadly he was deemed too sick to risk it and died never having tried it.

As David grew bigger so, necessarily, did the bubble, but even an upgraded bubble was still a small world for a growing mind. To help address this, a transport bubble was engineered to ferry him between the hospital and another bubble at home. His healthy elder sister (as a girl she was spared from suffering this type of SCID) would sleep in the living room next to his bubble, and friends could come round for a sleepover.

The closest David got to escaping from his bubble came courtesy of NASA: a $50,000 (£38,000) sterile spacesuit. The suit was connected to a sterile air-supply and allowed David to move in a world without walls for the first time in his life. He ran around in his garden and used a watering hose to drench delighted onlookers. Yet, amazing as the suit was, David only ever used it six times before outgrowing it. The suit required three people to get him into it safely (his parents and a trained NASA engineer) and the short battery life on the life-support unit only permitted an hour of actual usage once the cumbersome dressing process was done.

In the 1980s David approached his teenage years and it became increasingly apparent that the bubble could support his existence, but that risks would need to be taken if he were to really *live*. David's doctors told the Vetters that new research suggested that a less than perfectly matched bone marrow transplant could work. After deliberation with David about the risks, it was agreed they'd try a bone marrow transplant from his sister Katherine. David's mother has since reported that the doctors advised her the procedure was 99 per cent safe. This conflicts with the reality that the new research the doctors had at their disposal involved just two transplants: one that succeeded, and one that failed.

David's transplant was doomed to fail. Katherine's bone marrow was a Trojan horse, harbouring a hidden danger: the Epstein-Barr virus (EBV). EBV is familiar to most of us for causing the relatively harmless but socially awkward 'kissing disease' (also known as glandular fever, infectious mononucleosis or just plain 'mono'). For the majority of people the infection causes fever, fatigue, a sore throat and swollen glands in the neck. With a healthy immune system in place, the body triumphs over the virus and the symptoms clear up in two to three weeks. However, David's extreme immunodeficiency left him unprotected and vulnerable to a very unusual complication of EBV called Burkitt's lymphoma. This exceedingly rare form of cancer ravaged David's body, making him sick for the very first time in his life. The disease progressed quickly and less than four months after the bone marrow transplant, it became impossible to treat David inside the bubble. On 7 February 1984, the decision was taken to remove him from his protective bubble world for the first time in over 12 years. There was no going back and on 22 February, shortly after his very first skin-to-skin kiss from his mum, David passed away.

Despite the extraordinary lengths that David, his family, his medical team and even NASA went to, 12 years wasn't long enough to unlock the mystery of SCID. It would be another decade after David's death before scientists would finally pry open the Pandora's box of genetic mutations and identify which minor typo in his genetic code was responsible for his life-altering condition. It turned out to be a mutation in the IL2RG gene which sits on the X chromosome, which is why this variant of SCID is called X-linked SCID (XSCID). As we found in Chapter 6, boys are more susceptible to XSCID than girls, who are protected to some extent by having two copies of the X chromosome, hopefully thereby having one normal copy of IL2RG. However, this doesn't mean girls are immune to SCID; IL2RG is not the only typo that can cause the disease. In fact XSCID is responsible for just under half of all cases of Severe Combined Immuno-deficiency, and the other half are caused by a constellation of other mutations on different chromosomes. Expanding our knowledge of the litany of mutations that can cause SCID has helped us understand which genes make the normal immune system work. It has also revolutionised our approach to SCID, enabling the creation of new tests and treatments, some of which may sound like science fiction but are very much science fact.

Over a million babies across 16 states in the US have now been screened for SCID using a blood test that looks for small circles of DNA produced during the development of T cells. These circles, known as TRECs, are either absent or found only in tiny numbers in babies with SCID. Hunting for TRECs is a much better way of testing for SCID than simply counting T cells, because T cells from mum can also be found in newborns, confusing the results of the count. The TREC test also has the added benefit of using blood that is already collected in a tiny heel-prick test that is performed to screen for other serious conditions in new babies. This test can save lives. If SCID is caught early, before the baby has developed any infections, a matched bone marrow transplant offers a genuine cure for SCID.

Gene therapy

TRECs are a baby step compared to our next development. Leaping beyond the tried and tested and into the future, one of the hopes of SCID patients and researchers is that bone marrow transplants for SCID will become a thing of the past. The *real* payoff from the genetic revolution would be to perfect gene therapy. The idea is simple, deceptively so, and goes like this: if the problem is that children with SCID lack a normal copy of a gene, and we know what that gene should look like, why don't we just give them a normal one?

In 1990 a pioneering study in the US was the first to do just that, when four-year-old Ashanti DeSilva became the first person on the planet to receive gene therapy. Ashanti had SCID caused by a deficiency in an enzyme called ADA and the treatment aimed to give her cells enough working copies of the gene to make the enzyme from scratch. The result was a partial success – Ashanti is alive and well today, but her gene therapy wasn't a long-term cure so she still takes a synthetic tablet form of the enzyme too.

The quest for a genetic cure continued, and in 1999 a team of intrepid scientists in France attempted to give children with X-linked SCID a working IL2RG gene. To be able to get the gene into the bone marrow they harnessed the skills of a type of virus well versed in the dark arts of integrating into human DNA, called a retrovirus. This type of virus needs to insert itself into our DNA to be able to make its own proteins, however, the scientists were able to modify the virus so that instead of integrating and forcing the cell to make viral proteins and enzymes, it would cause the infected cell to make large amounts of the protein coded by the IL2RG gene. Using this viral equivalent of FedEx, the researchers were able to deliver a working copy of IL2RG into some cells harvested from the children's bone marrow. These cells were then transplanted back into the children, allowing them to make working T cells, B cells and NK cells for the first time in their lives. Perhaps the most incredible fact was that it worked and children with SCID were cured. The scientists had boldly gone where only science fiction had gone before, and made a gene therapy cure a successful reality. This wasn't just a breakthrough for SCID, it created hope for people across the globe with a plethora of genetic conditions, bringing the promise that the beginning of the end of genetic diseases could be nigh.

Sounds too good to be true? Sadly it was. The gene therapy came with a catch, and it was a major one. Shortly after announcing the success of the trial, the researchers discovered that two of the ten children cured of XSCID had gone on to develop leukaemia. In a 2003 article in the journal *Science*, the research team suggested the problem stemmed from the retrovirus they had used to introduce the IL2RG gene. By unlucky chance, in some of the bone marrow cells the retrovirus had inserted itself into the DNA near a gene called LMO2, cranking up its activity in the process. This was a disaster because the overactivity of LMO2 effectively took the brakes off the normal control process that regulates the number of new cells that are made. Without these brakes there was an explosion in the number of T cells, which meant that the children developed leukaemia. What had been recognised as a theoretical risk was suddenly a clear and present danger, causing scientists, patients and families to re-evaluate the safety of gene therapy.

A team of some of the world's best and brightest gene therapy experts were assembled from France, the UK and the US and they came up with a new plan: self-inactivating (SIN) viral vectors. This new generation of gene delivery vehicles were stripped of their superpowers of amplification, so they could deliver a gene but lacked the ability to massively ramp up the gene's capacity to make proteins. Crucially, this meant SIN vectors shouldn't be able to cause the increased activity of nearby genes in the way the original retrovirus vectors had. In theory. The only way to be sure was to test it. Nine children with XSCID from Europe and the US were enrolled in a trial to test the safety and effectiveness of the SIN system. In October 2014, the results of the study were published in one of the world's most prestigious journals, the *New England Journal of Medicine*, and the news was good. After between 12 and 38.7 months of follow-up, seven of the nine children had working T cells capable of killing infections. The fact that none of them has developed leukaemia is obviously fantastic, but the story isn't over and the team will follow the children's progress with intense interest, because the long-term effects of SIN gene therapy still aren't fully understood. It simply hasn't been around long enough yet.

So why would a parent sign their precious baby up to an experimental treatment? Well, while gene therapy carries risks, these risks have to be weighed up against the alternatives. Doing nothing is not an option. The current best alternative to gene therapy is a bone marrow transplant, which comes with its own risks and the very real possibility of failure. A European study that looked at 475 children with SCID from 1968 to 1999 found that among children who received a transplant from a matched sibling, 80 per cent were alive after three years. For a matched but unrelated donor, the odds were a little worse at 70 per cent and for children who received an unmatched transplant the three-year survival rate plummeted to 50 per cent. In the light of these odds, you can understand why mums and dads are willing to sign their children up to these trials and take on the terrible risks and tremendous potential benefits of gene therapy.

Leukaemia

All stories of immunodeficiency involve extremes: extremes of risk, extremes of science and extremes of numbers. Children with XSCID can have zero T cells, making their inability to fight off infection easy to understand. What's less intuitive is how a disease that involves having billions *more* white blood cells than normal also leaves people unable to fight infection. This is the situation people with leukaemia find themselves in. Leukaemia is a form of cancer that affects the cells of the bone marrow, causing them to multiply out of control. There are four main types of leukaemia based on the combination of the type of cell they affect *and* the speed of their progression. In terms of speed, 'acute' leukaemia tends to develop and progress rapidly, whereas 'chronic' leukaemia is usually slower to progress and more persistent in nature. When it comes to types of cell, lymphoid stem cells spawn all of our T cells, B cells and Natural Killer cells. A cancer of these stem cells produces lymphoblastic/lymphocytic leukaemias. Myeloid stem cells, on the other hand, would usually give rise to red blood cells, neutrophils, macrophages and other myeloid cells. A cancer of myeloid stem cells is called myeloid leukaemia. Thus our four main types are Acute Lymphoblastic Leukaemia, Acute Myeloid Leukaemia, Chronic Lymphocytic Leukaemia and Chronic Myeloid Leukaemia.

People with Acute Lymphoblastic Leukaemia (ALL) have a rapid rise in their lymphoblasts, an immature precursor cell to T cells and B cells. These immature cells are unable to fight infection, and they crowd out the bone marrow, turning the constrained space of the bones into something like a packed commuter train, leaving little space for any normal cells to join the carriage. This crowding out of the normal cell population is what causes the symptoms of leukaemia. The number of red blood cells and platelets drops, resulting in the easy bruising, ghostly pallor and exhaustion which typify ALL. Meanwhile the crash in working white blood cell numbers leaves the person vulnerable to all manner of infections, from the common cold to rare fungal infections. For reasons that aren't well understood, 85 per cent of cases of ALL occur in children under the age of 15, with most of those affected being two to five years old. Thankfully while children are disproportionately affected by ALL, they also have big numbers on their side when it comes to survival: 85 per cent of children can be completely cured of their ALL and go on to live long lives.

This statistic would have staggered Sidney Farber, the man regarded as the father of modern chemotherapy and a radical pioneer in the treatment of childhood leukaemia. His breakthrough came in the middle of the last century, when Farber was working as pathologist at a Boston children's hospital. It may seem odd that a pathologist would change the face of cancer treatment, but in the mid-to-late 1940s there was painfully little any paediatrician or oncologist could do for a child with leukaemia. Our understanding of the disease was woefully inadequate and, with the exception of a handful of rare spontaneous remissions, it was unfailingly fatal within weeks to months of diagnosis. Doctors like to cure things. It seems obvious, but it's worth saying aloud to help explain why nobody else was ploughing their energy into leukaemia research: they thought it was utterly futile. Farber had completed autopsies on over 200 children with 'acute leukaemia' but unlike many of his colleagues he refused to accept the incurability of cancer, and began a series of trials to find new treatments.

Today the testing of new medications for children is subject to an extremely rigorous ethics process, which means there is a long approval process to be navigated between identifying a drug and testing it in children. However in the 1940s the process was much less well developed, a fact that contributed to triumphs and tragedies. In one of Farber's early series of experiments, children with acute leukaemia were treated with folic acid. However, rather than slowing the progress of their disease as Farber had hoped, the drug seemed to accelerate the leukaemia and hasten the children's deaths. While their deaths were almost inevitable with a diagnosis of acute leukaemia, it was a sad reminder of the dangerous nature of trialling experimental drugs.

Yet one glimpse of a silver lining was salvaged from this disastrous experiment in the shape of an idea: if the presence of folic acid can accelerate leukaemia, can some sort of anti-folic acid slow it? It was a beautifully simple notion that changed the course of medical history because the answer, tentatively, was yes. In a triumphant example of bench-to-bedside progress, Farber commissioned the production of a purified anti-folic acid drug called aminopterin, and within five months of making it had treated 16 children. By early summer the results were in, and Farber published a paper in June 1948 describing them. Six did 'not respond well', with two failing to improve and four sadly dying. However, ten showed evidence of an 'improvement of important nature' in their disease, with blood counts shifting closer to normal and one child able to return to school. What was striking was not just the fact that they were improving, but the incredible pace of change – seriously sick children went from hospitalised to virtually normal in a few short weeks. The report of the findings emphasised the temporary nature of the treatment (one little girl relapsed after 47 drug-free days) and noted it had serious side-effects, including severe inflammation of the mouth. This well-balanced objectivity carried through to the paper's conclusions, where Farber's summation stated that aminopterin wasn't a cure for acute leukaemia but that it did represent a 'promising direction for further research'. Researchers are often criticised for overstating the nature of their findings. Contrastingly, this little summary was like taking a running jump off the other end of the scale, into the infrequently explored sea of scientific modesty. Make no mistake, aminopterin was one of the earliest chemotherapy agents known to medicine, and this was the very first use of a class of chemotherapy agent still being used by world-leading hospitals in cutting-edge chemotherapy regimes today. Sidney Farber gave the world a game-changer in the fight against cancer.

Immunocompromise

Another man who knows much about both drugs and cancer is Lance Armstrong. In 2013, almost two decades after overcoming testicular cancer, the seven-time winner of the Tour de France revealed he had used a banned substance called EPO (Erythropoietin). This drug was used by Armstrong and others to increase the number of red blood cells pumping round their veins, thereby increasing the oxygen-carrying capacity of their blood and so their exercise abilities. However EPO isn't the only reason for having more red blood cells than average. In 2003, British cyclist Charles Wegelius was temporarily banned from cycling for using EPO. He denied having touched the drug, and after further tests it became clear it wasn't that something had been added to Wegelius' body, it was that something was missing: his spleen.

The spleen is a multi-talented organ: it removes old red blood cells and plays a vital role in the immune system.[*] It's the ability to remove red blood cells that is salient to our cycling story. Five years prior to his ban, Wegelius had been in a quad bike accident which had damaged his spleen, necessitating its removal. Without a spleen to remove the old cells, Wegelius had blood with a higher than average red blood cell count, mimicking the effects of EPO. These cells weren't completely normal though. Before new red blood cells leave the bone marrow they have all but a tiny bit of

---

[*] And can even be used to predict the weather, at least according to a Canadian man who uses pig spleens to give a month-by-month annual weather forecast. For instance in 2015, 'The spleen shows a mild start to February and snow may disappear.' The forecast is limited to Canada so the rest of us will have to make do with less entrail-based systems.

their nucleus removed so that the cell can be packed tight with haemoglobin and carry as much oxygen round the body as possible. The spleen is responsible for removing the last little remnant of the nucleus, like stoning a cherry. If there's no spleen to do this, the little remnant stays inside the red cell and is visible through a microscope as a little pip called a Howell-Jolly body. So a simple blood test can paint a picture of whether someone has a working spleen.

Most of us have just a single fist-sized spleen that sits on the left-hand side of our body snuggled under the 9th to 11th ribs. Very rarely though, something goes awry in the earliest stages of embryo development and the spleen ends up in the wrong place. This isn't necessarily as bad as it sounds; in some cases it's like having a stranger unpacking your groceries and putting them in the wrong cupboards – it's not quite right but it's not *actually* a problem. This is often the way for people with situs inversus, who have all of their organs swapped to the opposite side of their body in an inverted image of normal without much impact on organ function. Others have a more higgledy-piggledy set-up, with some organs where you'd expect them to be and others not; this is known as situs ambiguous. There are varying estimates of how common a phenomenon situs ambiguous is. Rose and colleagues suggested an overall rate of at least 1 case per 40,000 live births; however, Gatrad et al. calculated 1 in 24,000 in an English population and 1 in 2,700 in a very inbred Asian population. Unlike situs inversus, at least 50 per cent of people with situs ambiguous have heart malformations as a result, which can be fatal or require significant corrective surgery. The heart isn't the only thing that's abnormal in situs ambiguous, and one way of classifying the various organ arrangements is according to the number of spleens the person has – they can be polysplenic (more than one spleen) or asplenic (no spleen).

The polysplenic have been found to have up to 16 spleens, though obviously not all full-size or there'd be little space for the other accoutrements of the abdomen, like the liver, bowel and bladder. Instead multiple little lobes create something like an elongated bunch of grapes for a spleen. In some people the little spleens twist on the stalks of the 'grape bunch', cutting off their own blood supply and dying. However, they can collectively create normal splenic function, which is more than can be said for people who are asplenic. Without a spleen they are immunocompromised, not as severely as someone with leukaemia or SCID, but much more at risk than the average person.

This is true no matter how the spleen was lost – whether it was removed by a surgeon following a quad bike accident or if it was simply never there in the first place. The splenically impaired are especially vulnerable to infections caused by encapsulated bacteria like *Streptococcus pneumoniae*, *Haemophilus influenzae* and *Neisseria meningitidis*. These bacteria have a sugar-based polysaccharide outer shell, which acts like a cloaking device that limits the activation of complement and therefore the body's ability to destroy the bacteria. This shell means the bacteria are far from sweet, and kill over 50 per cent of people without a spleen who are unlucky enough to catch them. Our understanding of why the capsule makes them especially dangerous for those without a working spleen is still emerging; clearly there are some cells in the spleen with a specialised killer skill-set that we rely upon to get rid of the crafty, cloaked bacteria. One possibility is the humble splenic macrophage, which studies have suggested plays a key role in eliminating encapsulated bacteria. However, the evidence is a little conflicted because mice with low numbers of these macrophages seem to be able to mount a response. An alternative nominee for the unsung splenic hero is a specialised type of B cell that's found in high numbers in the spleen. While the question of why the spleen is key is still being figured out, what we do know is that in its absence we have to intervene to help protect people against these infections. This means education, vaccinations and, for some people, life-long antibiotics. With all of this it's still possible to live a relatively normal life; after all Charles Wegelius went on to cycle in the World Championships, the Commonwealth Games and the Olympics, all *sans* spleen.

Do you believe in miracles? I don't, but I can see why the protagonist in our final tale of immunological vulnerability might. In the 1960s Ms X, as we'll call her, was one of the first people on the planet to be diagnosed with WHIM syndrome.* This condition involves recurrent infections, a plague of warts and an increased risk of cancer, all caused by a little mutation in the CXCR4 gene. This gene is crucial for allowing immune cells, especially neutrophils, to exit the bone marrow into the bloodstream. The WHIM-causing mutation in the DNA breaks this mechanism, leaving the cells trapped in the bone marrow and unable to do their job. This makes the patient vulnerable to all sorts of infections including the human papilloma virus, which causes the warts and the increased cancer risk.

More than 50 years after her original diagnosis, Ms X sought out specialist doctors at the US National Institute of Allergy and Infectious Diseases to get her daughters tested for WHIM. However, it was what she casually mentioned about her own disease that made the doctors' eyes widen: she hadn't had symptoms for almost 30 years. This spontaneous resolution of WHIM had never been reported before in the history of medicine and so the hunt was on to find the reason for her cure. In a solution so unlikely it was worthy of a made-for-TV movie, they found that Ms X had won the genetic lottery. A chromosome in one of her cells had shattered, a phenomenon known as chromothripsis, and it had happened to break precisely at a point that included the mutated CXCR4 gene in a stem cell that made neutrophils. Had this happened in another cell type it would have been completely inconsequential, but by affecting that one cell in that particular location, the chromothripsis created a stem

---

* WHIM stands for the main characteristics of the condition: warts, hypogammaglobulinemia, infections and myelokathexis. Hypogammaglobulinemia means the sufferer has low levels of gammma globulins and therefore antibodies in the blood, and myelokathexis indicates the retention of neutrophils in the bone marrow.

cell without the mutation that was able to produce neutrophils capable of escaping the bone marrow.

Most people with an inherited predisposition towards cancer are not this lucky. In fact cancer is one of the immune system's most difficult opponents. The reason for this is that spotting cancer cells is like playing the most difficult and deadly game of *Where's Wally?* Unlike spotting bacteria or viruses, which look very different from human cells, with cancer the immune system has to find Wally in a sea of a billion bespectacled geeks all wearing red and white striped jumpers and jeans, but with *ever* so slightly different smirks. That's the challenge of spotting cancer. That's our next chapter.

Cancer: A Deadly 'Spot the Difference'

Cancer has many faces, none of them pretty and many of them deadly. It wears a mask that conceals its killer features from its would-be assassins – the T cells and Natural Killer cells that roam our bodies. Despite its best attempts to see cancer's true colours, the immune system has a huge challenge on its hands. It's like trying to spot a murderer at the world's biggest masquerade ball where everyone's wearing the same mask, except that the homicidal guest's mask has ever so slightly different shaped pieces of glitter round the edges. Part of the reason a cancer cell's mask is incredibly convincing is that once upon a time it was just another average Joe.

Cancer is a multistep process. Discuss.

There are many things that can cause mutations in a cell's DNA to transform it from an average Joe into a cancerous Joe. From leather dust to Chinese-style salted fish, carcinogens make cells cancerous by endowing them with a very particular set of skills. In 2000, in one of the most widely read and shared scientific papers in the history of the top-tier journal *Cell*, Douglas Hanahan and Robert Weinberg outlined these skills in what they called the 'hallmarks of cancer'. The first is self-sufficiency in growth signals: while ordinary cells depend on receiving instructions to multiply, cancer cells are above such things and like malevolent rabbits on a reproductive rampage they replicate at will. Second, immunity to anti-growth signals: this is the yin to skill number one's yang, and essentially means that cancer cells ignore instructions to stop multiplying. Third, evading cell suicide: when a cell starts behaving abnormally, it can trigger an inner self-destruct mechanism called apoptosis. One of the main genes responsible for this self-destruct mechanism is called TP53, and if this gene is mutated in a way that stops it working, the cell can misbehave with relative impunity. It's recently been suggested TP53 may hold the key to a long-held conundrum: why don't elephants get more cancer? In the 1970s, researcher Richard Peto noted that there seemed to be no relationship between an animal's size and cancer rates. This was surprising because you'd expect that the cells of large animals would have divided more and therefore be more likely to have accumulated mutations and so develop cancer. However two studies published in October 2015 reported that elephants have 20 copies of TP53, where most mammals, including humans, have just one. Carrying so many spares offers elephant cells a fail-safe of epic proportions that should make it much less likely that an elephant cell would lose the TP53 self-destruct button and become cancerous. This isn't to say elephants are immune to cancer; as one scientist pointed out, if the elephant community took up smoking and had a bad diet we'd probably be seeing a lot more elephants with cancer.

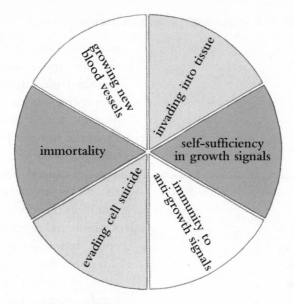

Figure 14.1 Hallmarks of cancer.

The fourth hallmark of cancer is immortality. Even once cells are able to grow despite the chemical melting pot they're stewing in telling them to cease and desist, they will eventually stop replicating. This is because cells have their own inner counter that puts an upper limit on the number of times they can divide before they die, thought to be 60–70 in human cells. Once cancer cells are freed from the tyranny of this counter they effectively become immortal. The final two hallmarks are growing new blood vessels, which ensures the growing tumour has an adequate supply of oxygen and nutrients, and invading into tissue to spread to other bits of the body.

While many of the risk factors that cause the mutations that produce the six hallmarks of cancer come from the things we consume or inhale, there is one important risk factor that comes from within: inflammation. Several diseases that involve chronic inflammation are linked to an increased risk of cancer. For instance, patients with inflammatory bowel disease have a five- to sevenfold increased risk of bowel cancer, and those with hepatitis C or B have a 20-times higher risk of liver cancer. In some ways this seems counterintuitive – surely having a hotspot of immune system activity should decrease the chance of a cancer's success? It's a bit like recruiting more police only to find your crime rate increases.

One possible explanation is that an inflamed bit of skin or gut isn't able to function as a normal barrier and instead allows environmental carcinogens to seep through and work their mutating magic on the cells they reach. Another possible explanation is that several of the chemicals involved in inflammation feed tumours. Solid tumours[*] can grow very quickly and when they get fat fast, the core of the tumour is often left a long way from a good blood supply. The resulting lack of oxygen causes the tumour cells to die, releasing their innards in a messy death that beckons white blood cells to the scene. The white blood cells set about secreting chemicals that promote the growth of cells and the development of new blood vessels. This is normally a great plan when bits of the body have been damaged, but in the case of cancer having these chemicals around is like bringing rocket fuel to a forest fire. These inflammatory chemicals are so beneficial to tumour survival that some tumours don't wait for the immune system to bring fuel to the party – they ooze their own versions of these pro-tumour chemical messengers into their environment.

At this point it may appear the activities of the immune system do cancer more good than harm. However in 2011, Hanahan and Weinberg added two new emerging skills to the list, one of which was evading immune destruction. According to one theory, known as immunoediting, the cancers that we see in people are just the tip of a malignant iceberg shaped and moulded by the immune system, which wipes out lots of cancers before we ever knew they existed. Thus it's not the case that all cancerous cells are able to outwit

---

[*] Solid tumours are exactly what they sound like – a solid mass, usually without any hollow cavities or liquid areas in it.

the immune system, but rather that we only ever see the ones that can. It's thought that one of the key steps in a cancer surviving the immune system is perfecting its mask of normality – either by losing markers on its surface that betray its cancerous ways, or by losing the protein platforms that show these markers to immune cells.

The immune system not only sculpts the face of cancer cells, it also influences which types of cancer we develop. For instance, people infected with HIV have a weakened immune system, and are 70 times more likely to develop non-Hodgkin's lymphoma and several thousand times more likely to develop Kaposi's sarcoma than the general population. Both of these cancers are known as AIDS-defining cancers as they mark the transition from being infected with HIV to having AIDS. Yet there are lots of other cancers that people with HIV are at no increased risk for, such as breast, colorectal and prostate cancer. The reason certain cancers are more likely to affect those with weakened immune systems is that they're caused by viruses – Kaposi's sarcoma by a herpes virus, non-Hodgkin's lymphoma by the Epstein-Barr virus – and people with HIV are less able to fight off these viruses.

In 2015, this well-known phenomenon took an extraordinary and unexpected twist as it was revealed a man had been diagnosed with the world's first tapeworm tumour. About 75 million people worldwide are thought to carry the dwarf tapeworm in their small intestine, where it lives a fairly innocuous life and causes its host few if any symptoms. In people with a compromised immune system, such as those with HIV, the worms can take up residence for years and accumulate in large numbers but they are rarely found anywhere other than the intestine. When a 41-year-old man with HIV was seen by doctors they probably thought the dwarf tapeworm was the least of his problems. He came to them with fever, cough, weight loss and fatigue, and blood tests revealed he had a CD4 count (a measure of a type of T cells) of 28. The normal range in a healthy adult is 500–1,200 and in someone with HIV a count of less than 200 implies progression to AIDS. A CT scan showed lumps in his lungs, liver and lymph nodes, so samples of tissue were taken from these and images sent to experts at the US Centers for Disease Control and Prevention for a telediagnosis. What they saw was perplexing – cells that seemed to have much in common with typical cancer cells but looked too small to be human. One possibility was infection with a plasmodial slime mould, but the truth was even stranger: the lumps had worm DNA in them. The doctors concluded that the worm's stem cells had been able to multiply in their immunocompromised host and over time had acquired the mutations that transformed them into cancer cells. Prior to this case, nobody had seen a tapeworm develop cancer, let alone seen multiplying, mutated tapeworm tumours inside a person. Sadly, by the time this diagnosis was made the patient was critically ill and opted for palliative care, dying 72 hours later. While we think this is incredibly rare, given the overlap between communities at risk of the dwarf tapeworm and HIV, it's possible worm tumours are under-diagnosed. The knowledge that this can happen should encourage the investigation of treatments, and also has the potential to spark a deeper understanding of the relationship between infections, the immune system and cancer.

Cancer drugs – something old, something new

The connection between inflammation and cancer also raises lots of interesting possibilities for novel treatments. The idea of new cancer treatments tends to conjure up images of cutting-edge experimental science (which we'll explore at length in Chapter 16), but some seeking the next cancer wonder drug are looking at something distinctly pedestrian: aspirin. It's over 2,000 years since Hippocrates described the pain-relieving properties of willow bark, and more than 100 years since its active ingredient was purified and turned into tablet form to be sold by pharmaceutical company Bayer. Today aspirin is one of the most widely used anti-inflammatory drugs on the market with an estimated 100 billion tablets swallowed across the world annually, and is used in the treatment of heart attacks and strokes. And perhaps tomorrow it will become the next big thing in cancer treatment.

You'd be forgiven for questioning whether the humble white pills so easily picked up in your local store could really deliver such grandiose results. Yet in September 2015, Dutch researchers announced the results of a study that observed nearly 14,000 cancer patients, some of whom had taken aspirin before their cancer diagnosis, some after and some of whom had never taken it. Those who took aspirin after a diagnosis of a gastrointestinal cancer were twice as likely to survive as those who did not take it. The result held even when the researchers adjusted for key factors that might have influenced patient survival, including the patient's age and sex, stage of cancer, treatment method, and other medical disorders. While this is clearly an interesting result, what's less clear is how aspirin could be having such an impressive impact. The scientists involved suggested the positive result could be due to aspirin's anti-platelet effect – the same talent that makes it useful in a heart attack. It's believed tumour cells bobbing about in the bloodstream try to evade the immune system by coating themselves in platelets, the little cell fragments that are essential to our ability to form blood clots. By interfering with how the platelets work, it's possible that aspirin ruins the cancer's ability to adorn itself with a platelet shield, leaving it exposed to the immune system. Alternatively, it's possible that aspirin's anti-inflammatory properties inhibit the immune system from creating the pro-cell-division soup we mentioned earlier, thereby inhibiting the cancer's growth. Whatever the mechanism, the prospect of a cheap, effective anti-cancer drug that's available in both resource-rich and resource-poor settings, with relatively few side-effects, is exciting many a scientist, and a cascade of research studies will soon be emerging. Priorities include understanding the appropriate dose and duration of aspirin therapy, as well as trying to identify whether it's possible to tell which tumour types could benefit most.

Yet maybe we shouldn't be worrying about details like tumour types, because it's possible popping an aspirin a day could be of value to a far wider range of people. A 2014 review of relevant studies by a team of researchers from institutions across the US and Europe concluded that there was 'overwhelming evidence' that regular aspirin use reduces people's chance of developing colorectal cancer. This was based in part on 20 years of follow-up from two trials which established a 37 per cent reduction in colorectal cancer in those taking high-dose aspirin for five years or more, although this result only became apparent a decade after the participants in the study had been randomly allocated to aspirin. There was also good evidence that aspirin reduced the risk of stomach or oesophageal cancer, and possibly breast, prostate and lung cancer too. Of course, as is often the case in medicine, aspirin is a double-edged sword, bringing with it some potential harms, from bleeding to stomach ulcers. Moreover, a few of the study's authors declared relationships with pharmaceutical companies, including those with an interest in anti-platelet drugs like aspirin. So we're not quite at the stage of unequivocally recommending an aspirin a day to keep the doctor away. Nevertheless, the authors suggested that for the average 50–65-year-old, taking aspirin for a decade would reduce their relative risk of cancer, heart attack or stroke by 7 per cent in women and 9 per cent in men over a 15-year period. Over 20 years they suggest aspirin is associated with a drop in the relative risk of death of 4 per cent across 20 years – that equates to 130,000 lives that could be saved in the UK alone. All with a drug that costs less than 2p (2.6¢) per tablet.

Aspirin isn't the only drug with the credentials to be repurposed as a cancer drug. The Bacillus Calmette-Guérin (BCG) vaccine is not a high achiever in the ranks of modern vaccines. Despite having formed part of a routine vaccination programme against tuberculosis (TB) in multiple countries for years, it is largely ineffectual in stopping the spread of TB. Estimates of its effectiveness against TB range from 30 to 66 per cent and the trials where it demonstrated effectiveness were largely in childhood, when TB is rarely contagious. Reflecting the scientific community's ambivalence towards the vaccine, different countries adopted different approaches. In 2005, a study of all 25 countries in the EU found it was recommended for all children under one year old in twelve countries, in older children in five countries and in children at high risk in ten countries. BCG used to be given routinely to British schoolchildren but was abandoned in 2005 as research suggested that for every 5,000 children vaccinated, a single case of TB would be prevented in the subsequent 15 years.

However, while BCG failed to hit top marks as a vaccine to stop the spread of TB, it has since found its niche in a very unexpected place: the bladder. I have no idea how someone came to discover this, but if you fill the bladder with a BCG-containing solution it's very effective both at preventing cancer cells from invading deeper into the bladder and at stopping the cancer returning once it's been cleared. In fact BCG is the only treatment approved by the US Food and Drug Administration as the primary therapy for 'carcinoma in situ' – an early form of bladder cancer where the cancer cells are limited to the inner lining of the bladder. While it's been clear for over 30 years that BCG works, what's less clear is precisely how. We know it's a form of immunotherapy* because a functioning immune system is essential to its success, and the pee of post-BCG patients is rich in neutrophils and macrophages. The white blood cells are thought to destroy the cancer cells directly, and also secrete tumour-killing proteins like TRAIL (tumour necrosis factor-related apoptosis-inducing ligand), which is found in high concentrations in the urine of patients who respond to BCG treatment, but in low quantities in those who are non-responsive. Why BCG is so effective at triggering this response in this particular malignancy remains a mystery – it can't do it in any other cancer.

---

* Immunotherapy is a form of treatment that uses substances to enhance or suppress the immune system in order to help the body fight disease.

This fortunate mystery is not a unique occurrence – there are other micro-organisms out there capable of killing our cancers. A new bacteria-based cancer killing technology benefited Layla, one of the luckiest and unluckiest babies on the planet. At just three months old she developed acute lymphoblastic leukaemia (ALL), which, as we found in the last chapter, is a very serious but relatively rare cancer of the blood. About 85 per cent of children who develop ALL respond to treatment and are cured, but Layla was in the unfortunate 15 per cent. However, her doctors were able to offer her parents some hope in the form of a radical new treatment only tried once before in people and never in children with leukaemia: genome editing. This technique can use TALEN (transcription activator-like effector nuclease) enzymes from the primitive immune system of a plant-infecting bacterium to alter the DNA code of human cells with incredible precision. This allows scientists to play with the blueprint of cells and change the way they behave. In Layla's case, T cells from a donor had their DNA re-engineered so they would only attack the leukaemia cells in her blood, not her other body cells. The T cells were also made less vulnerable to the chemotherapy being used to kill Layla's leukaemia. The treatment wasn't a cure, but it had given her something precious: time. Time is the thing children like Layla need to find the matched bone marrow donor that will finally cure their illness. Following a successful bone marrow transplant Layla's doctors declared her free from cancer. Although there are no guarantees the cancer will remain absent in the long term, having a healthy little girl less than six months after being told palliative care was her only option was a better result than her parents or doctors could ever have hoped for.

The story of genome editing is still just being written, but as the chapters unfold through the boundary-pushing experiments of scientists and the early promise of patients like Layla, it has the potential to be one of the most revolutionary stories in the history of medicine. So many diseases are caused by genetic errors, from cancers to degenerative neurological conditions, and genome editing offers the prospect of applying molecular surgery to cure them all. Its potential benefits even extend to infectious diseases, with HIV the very first example of genome editing being applied to people. In 2014, the prestigious *New England Journal of Medicine* reported that researchers had treated 12 HIV-positive patients with a genome editing technique aimed at rendering their T cells immune to the virus. They extracted T cells from the patients' blood and, using a different type of enzyme from Layla's treatment, reprogrammed the T cells' DNA so they would no longer display one of the receptors the virus relies on to force its way into the cell. While the study was small, the results were impressive, with half of the patients able to stop their anti-HIV medication following the genome editing therapy. There is still much more work to be done, including following patients' progress in the long term and trying the treatment in far larger numbers of patients, but genome editing is a space everyone should be watching.

Cancer-killing viruses

Yet another example of a cutting-edge addition to our cancer-fighting arsenal based on skills borrowed from another species are the oncolytic viruses. Oncolytic (which translates roughly as 'cancer-disintegrating') viruses infect and kill cancer cells but not healthy cells, and can be naturally occurring or engineered in a lab specifically to prey on cancer. They work by infecting and then exploding out of cancer cells, thereby killing them, and also by drawing the attention of the immune system, which starts to take on both the virus and the tumour. There are lots of advantages to using these viruses instead of other treatments like chemotherapy. For instance, cancers can mutate in a way that makes them resistant to the attacking action of individual chemotherapy drugs. However this resistance is much less likely with the oncolytic viruses because while many drugs are one-attack-method-wonders, the viruses have a Swiss army knife selection of killer techniques at their disposal. They're also more selective cell killers than many drugs, and this high-specificity targeting of just the tumour means they are less toxic for patients. The fact that they multiply in the tumour also means that unlike drug levels, which fall off if not topped up, the cancer-killing power of the oncolytic viruses actually builds over time.

Of course there are challenges too, and for years the race to get an oncolytic virus available to patients was characterised less by scientific triumph and more by a series of failed early clinical trials. One of the key challenges was how to get the virus to the tumour without the immune system destroying it en route. After all, no competent immune system should allow even modified versions of the viruses being used – including measles, polio and herpes – to pass by without alarm. Various solutions are being trialled in labs across the planet, including modifying human cells to act like tiny virus-mules to ferry the virus through the blood to the tumour, or giving drugs alongside the virus to temporarily suppress the immune system.

While this work continues, a more direct solution has already proved successful: injecting the virus directly into the tumour. This isn't a solution for all cancers, because many are inaccessible, but it is a strategy that works in skin cancer. This is how it finally came to pass that in October 2015 the wait for the world's first oncolytic virus was over and the FDA approved an injectable form of T-VEC, a modified herpes virus, for the treatment of inoperable melanomas (skin cancer). In addition to being programmed to infect only tumour cells, T-VEC's modifications also endow the herpes virus with a gene for making a molecule called GM-CSF (granulocyte-macrophage colony-stimulating factor), which boosts the immune system's own ability to kill cancer cells. The study on which the FDA based their approval decision had 436 patients with metastatic melanoma, an advanced form of a particularly lethal skin cancer that kills 75–90 per cent of sufferers within five years of diagnosis. The study showed that 16.3 per cent of those receiving the virus had a decrease in the size of their skin and lymph node lesions for at least six months. This compared with 2.1 per cent in the group receiving standard treatments. This is a really positive result for a disease which has few effective treatments in the advanced stages and is on the rise – in the US the incidence of melanoma has doubled since 1973 and in England the melanoma rate is doubling every 10–20 years. Yet despite this positive result, and the undeniably pioneering nature of T-VEC, celebrations of this new therapy must be tempered by the fact that there was no overall improvement in the survival of the patients or indeed any effect on melanoma that had spread to internal organs like the brain, bones or liver. However there were also few side-effects and now T-VEC has blazed a trail through the regulatory red tape, other oncolytic viruses will undoubtedly follow, opening a whole new line of attack in our anti-cancer arsenal.

The term cancer covers a gamut of diseases, some of which have a very hopeful prognosis and others that are far more deadly, but all of which are formidable opponents for the immune system. In our next chapter we will meet two infections, Ebola and anthrax, and see what makes these infections as terrifying and as lethal as the most deadly cancers.

CHAPTER FIFTEEN

Killer Bugs: Be Afraid, Be Very Afraid

There are diseases that are simply annoying, like the common cold, and there are diseases with mildly amusing names, such as Jumping Frenchmen of Maine.* And then

---

* This very real condition was first discovered in Maine in the late 1800s in an isolated group of lumberjacks of French-Canadian descent who were all afflicted with an unusually extreme startle reflex. The cause of their condition, which is extremely rare and definitely not amusing for those with it, remains unknown to this day but leads to jumping, flailing, screaming and even throwing objects when startled. Despite the name, people with a very similar constellation of symptoms have been described all over the world, from Siberia to Somalia.

there are the diseases that inspire fear and legend: the microbiological mass murderers. This chapter is about the notorious few, the serial killers that have swept through countries, transforming cultures, wiping out entire generations and dictating history to the human race. We will travel to West Africa and the scene of the 2014 Ebola epidemic to see what makes one of the most lethal killers in modern times so different from the billions of other microbes we inhale and eliminate with ease. We'll also see how some people are hacking history's most deadly bugs to create their own perfectly lethal weapon.

An Ebola epidemic

In 2014 in the humid heat of Liberia's summer, Harrison Sakilla journeyed north across the country before crossing into Sierra Leone to care for his sick mother. He assumed she had been stricken with malaria, a common and sometimes deadly gift from the local mosquitoes. After several days of caring for her and watching her wither, Harrison's mother sadly passed away and he was left to provide one final act of love: he buried her. What Harrison didn't know was that his mother had something far more deadly than malaria. As he buried her body it was bursting with billions of invisible particles of one of the most dangerous viruses in history: Ebola. In the months to come, burials in Sierra Leone would change beyond recognition, with hazmat suits, bleach and burning replacing the tender farewell of a normal funeral. However, at the time Harrison was oblivious to the threat and travelled back home to Liberia to be with his wife and six children. It didn't take long for his world to tip upside down. It began gently at first, with diarrhoea and a headache, but within days his symptoms became unbearable and his life unrecognisable. Harrison took a journey into the unknown, leaving his village for somewhere nobody else in the country had been yet: he became the first patient admitted to an Ebola Treatment Unit in Liberia. Owing to the distance from his village, and the danger of his disease, Harrison had few visitors during his time at the centre and saw many people stumble in with the same symptoms, only for them to exit in a body bag. Luckily for Harrison, he may have been the first to walk in, but he was also the first to walk out.

Although Harrison's life was spared, it was far from untouched by death: he lost both of his parents, his sister, his niece and his great-niece to Ebola. Their names numbered among the thousands of men, women and children killed by the most deadly outbreak of the disease in recorded history. The world had got used to this being a relatively obscure disease limited to occasional small eruptions dotted across remote and rural patches of the African continent, and successfully confined by quarantine. Between these outbreaks the virus dropped off the public health radar entirely, and between 1979 and 1994 not a single case was recorded. On average the entire planet of several billion people might witness just 500 cases of Ebola in a whole year. This time was different. At the peak of the 2014 outbreak, the nation of Liberia alone reported 300–400 new cases in a *single week*. The cumulative result was that from the day of the first recorded case in December 2013 to mid-August 2015, the US Centers for Disease Control had been notified of over 28,000 cases of Ebola. Of these, 11,301 people died. That's a greater death toll than every previous outbreak over the last 40 years combined. While the West African countries of Liberia, Sierra Leone and Guinea were the hardest hit, the virus went globetrotting, with cases of the disease being recorded in 10 additional countries. It even extended its tendrils to the previously untouchable nations of the United States and the United Kingdom through infected aid workers.

The world was completely unprepared for the onslaught. Hospitals, charities, esteemed universities, governments, and even the supposed guardians of global health, the World Health Organisation, were all caught with their public health pants down, and were thoroughly spanked by the scale and speed of Ebola's spread. Due to the relative rarity of the disease and the fact that it tended to affect rural areas of low-income countries, understanding Ebola (let alone developing a cure for it) hadn't been a major science funding priority for drug companies or wealthy countries. The West African outbreak changed all that. The year 2014 saw the dawn of a new holy grail for global health: stopping Ebola. So began a scientific stampede to understand the inner workings of the virus and to uncover what set survivors like Harrison apart from the thousands that succumbed.

At the head of the pack were the Ebola hunters* – the doctors and scientists charged with tracking down 'Patient Zero' (the codename for the first person to contract the strain of Ebola responsible for the outbreak). They painstakingly traced the paths of those infected and fanned out the investigation to include the contacts of all those infected. It was like following the threads of a complex web to try and find the spider at the centre of it all. Eventually they found a thread that led them deep into the forests of Guinea to the point where the borders of Guinea, Sierra Leone and Liberia collide.

It was here they heard the story they had been seeking: on 2 December 2013 a two-year-old boy had fallen ill with a mysterious illness characterised by fever, vomiting and black bowel movements (a sign of internal bleeding). He became very ill very quickly, and sadly died just a few days later. But his death was only the beginning. He was swiftly followed by his three-year-old sister, his mother, his grandmother, a nurse and then the village midwife. At the time nobody knew what the devastating illness was, a fact that didn't generate as much alarm as you might expect because the doctors knew all too many killer diseases that could have been the culprit. The symptoms hinted at a hemorrhagic fever of some kind, but in medicine you're taught to expect the expected – to look for horses not zebras† – and Guinea hadn't seen a single case of Ebola in its history. It would have taken someone familiar with Ebola to make the mental leap to suggest it was to blame.

---

* Or in more scientific but less sexy parlance, epidemiologists.
† A phrase appropriate for the UK or US, but frankly misleading in other bits of the world.

When the international team of researchers ventured into the local hospitals to investigate the illness, they took blood samples from patients with symptoms very similar to those of the two-year-old. When they looked at the blood down the microscope they found conclusive proof of the first ever outbreak of the Ebola virus in Guinea. The results were published in the *New England Journal of Medicine* in an article that reflected the extraordinary nature of live field research during an emerging epidemic. Two phrases in the manuscript stand out: 'clinical data were not collected in a systematic fashion' and 'informed consent was not obtained'. Ordinarily these phrases could easily stop an article being published in any medical journal, yet this piece of research made it into one of the most prestigious journals in the world. The reason was simple: these were not ordinary times and the work wasn't a piece of meticulously calculated academia, it was part of a public health response to one of the biggest threats to human health in recent times. From the time of the realisation that an Ebola outbreak was underway, and throughout the dark months that followed, knowledge was power.

However, knowledge was in seriously limited supply. In fact, our knowledge of the existence of Ebola only stretches back to 1976 and begins with the death of a Belgian nun in Zaire (now the Democratic Republic of Congo). She had died of a strange illness, the like of which the local doctor (also a Belgian abroad) had not seen before. Curious, he put several vials of her blood in a coffee thermos, packed them tight with ice and shipped the precious cargo to the Institute of Tropical Medicine in Antwerp in the hand-luggage of a passenger on a commercial jet. Clearly airport security was *slightly* different then.

On the other end, an unsuspecting 27-year-old trainee microbiologist by the name of Peter Piot took delivery of the apparently unremarkable thermos. He found it filled with a few mainly melted ice cubes and some vials of blood, a few of which were broken and seeping their ruby contents into the melt water. Unperturbed, Peter and the team set about doing what they always did with random gifts of blood-borne mystery, and had a look at the blood under the microscope. What they saw stopped them in their tracks: a humongous virus. The closest thing any of them had ever seen before was Marburg virus – a deadly hemorrhagic fever discovered 10 years earlier among researchers in Germany working with African green monkeys. However, rapid checks with Marburg experts confirmed this was something new, something never witnessed by human eyes before. This titan among viruses would come to be known as the Ebola virus.

In 2014, almost 40 years after he first peered down a microscope at the virus, Peter Piot found himself facing Ebola once more, but from a slightly different vantage point. Peter the trainee microbiologist had become Professor Piot, director of the London School of Hygiene and Tropical Medicine, and one of the world leaders in the public health effort to stop the biggest Ebola outbreak in human history. This time round some of the mystery had already been unravelled, allowing researchers at least a little insight into Ebola's modus operandi. For instance, five different species of the Ebola virus had been discovered: *Zaire ebolavirus*, *Sudan ebolavirus*, *Tai Forest ebolavirus*, *Bundibugyo ebolavirus* and *Reston ebolavirus*. All except *Reston ebolavirus* (which is confined to animals in the Philippines) are known to cause disease in humans, and each has its own home turf across the African continent. The species responsible for the 2014 West African outbreak was *Zaire ebolavirus*, the most deadly of the five with a fatality rate of 55–90 per cent.

The *Zaire ebolavirus* genome consists of a single long strand of genetic material that codes for just eight proteins, which is all the virus needs to bring an entire health system to its knees. Scientists studying the secrets of this genetic code found that the virus mutates at a super-high rate, partly because of the type of genetic material it has, known as RNA, which has a very error-prone replication process. This ability to mutate rapidly makes the virus very adaptable, which is a concern during a long outbreak as every day increases the risk of the virus transforming into something even more deadly or transmissible. One spectre (raised mainly by hysterical media coverage) was the prospect of an airborne Ebola that could be passed on in aerosolised form when an infected person coughed or exhaled. At present the virus can only pass through broken skin or mucous membranes (like the eyes, nose and mouth) when people come into direct contact with infected body fluids, from blood to breastmilk, sweat to semen. The chances of Ebola becoming transmissible on the air are very, very small – it's exceedingly rare for a virus to change its method of transmission, as demonstrated by the fact even a mega-mutater like HIV hasn't undergone this scale of metamorphosis. Moreover, a study of the mutations in Ebola over multiple outbreaks reported that in 40 years not one mutation had made Ebola any more or less able to infect or kill people. Of course it was pretty good at both to start with.

While we've torn the lid off many of Ebola's past secrets, there is still so much about this killer that we suspect but have yet to prove beyond a reasonable doubt. In particular two pretty fundamental questions remain. First, in the times between recorded outbreaks, where is Ebola hiding? And secondly, Ebola outbreaks can kill up to 90 per cent of those infected – so who are the 10 per cent and why do they survive?

At present, science gets partial credit for answering the first question with 'fruit bats'. The reason it's only partial credit is because the live virus has never actually been found in the blood of wild bats. However, there is a lot of circumstantial evidence for this theory, like the fact that blood tests on bats have found antibodies to the virus, implying they've been infected in the past. The theory goes that the *Pteropodidae* family of fruit bats are the natural host of the Ebola virus, and that the virus makes the bats sick but isn't nearly as lethal as it is in people. The result is the bats are still able to continue travelling vast distances, carrying the virus as they fly across the African continent. This would explain how the *Zaire ebolavirus* made its way to West Africa when previously it had been restricted to three Central African countries. These countries have little trade with the Gueckedou area of Guinea, where the first recorded cases appeared. Moreover the remote nature of Gueckedou made it highly unlikely that a person infected in a Central African country would be well enough to make the whole journey. Thus migrating bats were a much more plausible mule for importing the virus to Guinea.

If we accept that bats spread Ebola, the next logical question is how the virus makes the leap to people. While this may not be an everyday occurrence in some bits of the world, bats and people regularly come together in rural West Africa, both in the forests and at the dinner table. Bats are a common type of bushmeat hunted as an important source of protein, and served either in a peppery soup or smoked over an open fire. Though it was unclear which bat body fluids might be Ebola-laden, the Guinean government took the precautionary approach of applying a blanket ban on the sale and consumption of bat soup and other bat-meat products during the Ebola outbreak to try and reduce the risk of transmission.

We know the family of the two-year-old who first contracted Ebola in the outbreak hunted bats, and that he fell sick after playing in a guano-laden tree hollow. However we'll never know if the bats were definitely to blame because, either by coincidence or design, the tree near his village that housed thousands of bats was burned down shortly after the local people were told of the Ebola outbreak. When German scientists arrived to investigate all they had to work with were a few charred fragments of bat DNA. But the quest to definitively prove whether bats are indeed where the virus hides between outbreaks goes on, and with more gusto (and funding) than ever before, so it seems likely Ebola's days of keeping this particular secret are numbered.

The secret of the 10 per cent survivors is rather more elusive. In the massive West African outbreak the odds of survival were better than 1 in 10, but what set Harrison apart and enabled him to join the ranks of the Ebola survivors remains a mystery. Was it something about his immune response to the disease? Or the medical treatment he received? Or some small genetic quirk that protected him? One of the biggest stumbling blocks to answering the question has been that most of what we know about Ebola has come from studies in guinea pigs, monkeys and mice. This is essentially because conducting rigorous trials is unsurprisingly low on the agenda of anyone desperately clamouring to save lives and stop an Ebola epidemic. Also, Ebola outbreaks don't tend to work on the same timescales as the average research funding or ethics application.[*]

While we still don't know what separates the survivors from the crowd, researchers have pieced together some of the ebola story using the patchwork of evidence at their disposal. The virus seems to infect lots of different cell types, but macrophages are probably among the first to be infected. As we know, these white blood cells roam the body, so when the replicating virus is finally ready to pop its little incubator open it could be unleashing the new virus almost anywhere in the body. This process spreads the virus far and wide, allowing it to infect everything from the skin to the liver. At the peak of the infection one-fifth of a teaspoon of an infected person's blood can contain 10 billion viral particles. The same volume of blood in an untreated person with HIV would have 50,000 to 100,000 particles. For the infected person, the consequence of this massive viral load is often massive diarrhoea and/or vomiting of up to 10 litres (21 pints) a day. What's not clear is whether the virus is causing these devastating symptoms directly, or if it's a result of the torrential release of immune messenger chemicals by infected macrophages tripping a massive uncoordinated immune reaction. This includes the rapid and severe bleeding which gave Ebola its original name of 'Ebola hemorrhagic fever'.[†] As the macrophages and other cells die, they also cause 'bystander' deaths in T and B cells, effectively crippling the targeted weaponry of the adaptive immune response, which would normally overwhelm an invading microbe. That's not to say the virus doesn't do direct damage too. It seems to prevent some infected immune cells from maturing, which stops the

---

[*] In truth not many things do, except maybe sloths, snails and glaciers.

[†] Something since dropped because major haemorrhage doesn't occur in the majority of cases.

innate immune system from sounding the alarm to the T and B cells. This explains why some people die from Ebola without ever managing to develop antibodies against it.

The only silver lining from the horrifically long duration and unprecedented size of the West African outbreak was that it allowed researchers to expand our knowledge of Ebola. For instance, researchers from Washington University School of Medicine discovered that one of the ways Ebola disrupts the immune system is by blocking an emergency fast lane in cells that would usually allow them to respond to viral infections rapidly. Other researchers focused on the tribe of the Ebola survivors. While they were shunned in fear by many of their friends and relatives, the blood of these survivors became a great source of fascination and hope for researchers. Anecdotal evidence implied that giving sufferers a transfusion of blood from a survivor could improve their survival chances, presumably by giving them antibodies against the virus. A small study from a 1995 outbreak reported that seven out of eight infected people who received a blood transfusion survived the virus, but the lack of a comparison group made it hard to judge if the blood was a cure or if they were just seven very lucky people.

Yet the real aspiration among those who witnessed the devastation of Ebola was not just to treat it, but to prevent it. The first major breakthrough in this quest came in August 2015 when a clinical trial of a new vaccine hit the headlines with some very promising early results. The researchers had given the vaccine to thousands of people in Guinea who had been in contact with someone with Ebola. This technique is known as ring vaccination and has good prior public health form, having previously been used to help eradicate smallpox. In the Ebola trial, half the volunteers were given the vaccine immediately and the other half were given it three weeks later. In those vaccinated immediately there were zero cases of Ebola, while in the group that had to wait for the vaccine 16 people (0.5 per cent) developed symptoms. This result was a cause for celebration, as it represented the most promising evidence to date that the world would soon have its first effective Ebola vaccine. It was also proof that with co-operation between funders, academics, governments and industry, high-quality scientific research could occur at the break-neck speed necessary to get a safe vaccine to people when they needed it most.

However, the lab science is just one part of taking on Ebola. Even if the vaccine fulfils its designers' wildest hopes and dreams in terms of effectiveness, we're still a very long way from watching Ebola slink off to join smallpox as an eradicated disease. First we need to take on far more unpredictable and difficult to study foes than the virus itself, namely human fear and the absence of trust.

Nine months after the outbreak in Guinea began, a small group of health workers and local journalists travelled to the remote village of Womey to raise awareness of Ebola. According to a witness the meeting began well – the chiefs welcomed the delegation with kola nuts and the conversation began. However things soon took a horrific twist. A group of young villagers came and started to stone the visiting staff, before dragging some of them away. When the staff failed to return from the visit, officials initially reported that they were being held captive and that residents of the village had destroyed access bridges to prevent any outsiders reaching them. When the authorities finally arrived in the village they found seven bodies in the latrines, three with their throats slit. This was clearly a terrible tragedy, and sadly not the last example of Ebola fighters losing their lives just doing their jobs. However it was also testimony to the level of fear and suspicion the Ebola epidemic had engendered. Ebola had never been seen in Guinea, and now, as foreigners and people clad from head to toe in alien white suits and masks suddenly descended, so did death. It doesn't take a massive leap of the imagination to see why some of the local people came to the conclusion that the outsiders claiming to stop Ebola were actually the ones bringing it in.

One of the particularly problematic symptoms of this suspicion and mistrust was the concealment of the dead, as families feared their loved ones would be taken and denied the rituals of burial – namely being washed by family members first. They were right. When people died from Ebola the authorities rushed in to douse the body in bleach, sealed it in a plastic bag and hurriedly buried it to prevent the virus being passed on post-mortem. When the bodies were instead washed in secret and buried under cover of darkness, the mourners were unknowingly placing themselves in grave danger. Aware that this behaviour was causing them to fall short of their targets for safe burials, the government in Sierra Leone responded by sending teams house-to-house to search for bodies and take them, deepening the distrust and distress in the community.

These were very human reactions to deeply inhumane circumstances. What is harder to understand, and what perhaps best demonstrates the power of Ebola phobia, is the reaction from those countries least affected. In 2014 the UK government was so mindful of the fears of its almost universally safe populace that they put in place a screening programme at airports with little to no evidence base and at great financial cost. The last time such a white elephant of a screening programme was implemented was in Canada during the SARS (severe acute respiratory syndrome) crisis. At that time, 677,494 people who flew into Canada completed a health questionnaire, of whom 2,478 answered 'yes' to at least one question. This triggered an assessment by a specially trained nurse, which included more in-depth questioning and having their temperature taken. None of them had SARS. To complement this system, the government installed thermal scanners at six major airports. Of the 467,870 people screened, 95 were referred to a nurse for further assessment and not a single one actually had a raised temperature. All of this came at the low, low price of $17m Canadian dollars (over £10 million). Based on the Canadian experience it would probably have been cheaper, and about as effective, to employ imaginary Ebola-sniffer cats at Heathrow.

Similarly, while the Ebola fighters were collectively named *TIME* magazine's person of the year 2014, the reception they received in their communities was often very different.

Foreign workers who returned home often received a frosty welcome, with some American staff ordered into quarantine despite a complete absence of evidence that they were infected. Although 80 per cent of the British public supported and admired British healthcare workers who had travelled to take on Ebola, 75 per cent said they would not shake the hand of someone who had recently returned from caring for Ebola patients. Times were even tougher for the locals who took on the thankless task of burying bodies (for about $100 or £80 a month). The 'burial-boys', as they were often called, were ostracised by their communities, leaving them alone with the mental horrors of a day's work. Healthcare staff also suffered from seeing their colleagues die around them. By the end of August 2015, 881 medical staff across Liberia, Guinea and Sierra Leone had contracted Ebola, with 512 of them dying. The USA has about 245 doctors per 100,000 people, while Guinea has just 10. This fact made the loss of life among medical staff not just tragic but also a threat to the country's ability to provide basic care even for non-Ebola patients. One group of researchers calculated that 74,000 fewer malaria cases were seen at health facilities in Guinea in 2014. Malaria hadn't gone anywhere, but the ability to treat it, and the willingness to seek medical care, had.

Bioterrorism

From macrophages spewing out haemorrhage-inducing chemicals, to the terror-inspired responses of populations and governments, a huge chunk of what makes Ebola such an accomplished serial killer is us. However, as we better understand the virus and develop effective treatments, we will also be Ebola's undoing. Perhaps one day we will consign it to the history books with other once horrifying scourges. Or at least that's what most people hope for. Yet while plague, smallpox and Ebola inspire fear in most of us, for some individuals they are the tool with which to realise their dreams and everyone else's nightmares. The US Centers for Disease Control lists six Category A priority pathogens – the 'organisms/biological agents that pose the highest risk to national security and public health'. Among the six are Ebola (as one of the viral hemorrhagic fevers), plague, smallpox and anthrax, which have made it onto the list because of a combination of four key traits. First, they can be either easily disseminated or easily spread from person to person. Second, they have high mortality rates and could have a significant impact on the health of a population. Third, they have the potential to cause public panic, as we saw with Ebola, and social disruption. Finally, dealing with these diseases is not an everyday event and therefore countries have to take special action to be prepared, such as stockpiling treatments and having emergency plans in place should the worst ever happen. In the twenty-first century the prospect of germ warfare using any of these pathogens tends to conjure up images of terrorists in gleaming white body suits concocting weaponised strains of deadly plagues in hidden high-tech labs. Yet the roots of germ warfare are exceedingly old and exceedingly dirty.

Mid-fourteenth-century Europe saw the dawn of the Black Death. This scourge swept the continent, killing 30–50 per cent of the population in five short, brutal years, leaving a blanket of corpses shrouding once vibrant cities like London and Paris. Dedicated plague cemeteries were opened in London with 200 bodies being buried every day, some piled five deep because of the massive pressure on space in the infected heart of the city. Rats and their fleas have widely been credited with spreading the disease, which was caused by the bacterium *Yersinia pestis.** We know it could also be transmitted between humans by contact with the infected

---

* There was some debate regarding which bacterium actually caused the Black Death, but that was laid to rest in 2011 by a group of researchers who weren't afraid to get their hands dirty. They exhumed the bodies of long-dead plague victims from the London plague cemeteries, extracted their teeth and drilled into them. What they found was *Yersinia pestis* DNA, 650 years after it had killed the souls filling the graves.

tissue or body fluids of a plague victim, and there are tales of humans giving the Black Death a helping hand.

In 1346 the Mongol army laid siege to the city of Caffa (now Feodosia, Ukraine), a cosmopolitan port that had been established by the Genoese to dock their vast merchant ships. While the Italians brought much trade to the area, their presence was founded on an uneasy alliance with the local Mongols. This alliance snapped in 1343, starting an on-again off-again siege, which in 1346 was very much on. That was until a plague began laying waste to the Mongol fighters as described in the memoirs of an Italian notary by the name of Gabriele de' Mussi. Thought to be a reliable, though second-hand, source, De' Mussi describes how the plague 'killed thousands upon thousands every day ... as though arrows were raining down from heaven to strike and crush the Tartars' arrogance'. So, a reliable but slightly biased account. He goes on to describe the swellings in the groin or armpit and the 'putrid fever' that typified the plague, and the helplessness of the medical profession to save the sick.

The people of Caffa had largely been quarantined from this horrific disease, a state of affairs the Mongol army apparently decided need rectifying. So they began catapulting corpses into the city. In one of the first, and most spectacular, recorded instances of germ warfare, De' Mussi describes how the 'rotting corpses tainted the air and poisoned the water supply, and the stench was so overwhelming that hardly one in several thousand was in a position to flee'. He suggests that those who did flee brought the plague with them to mainland Italy, acting as a key conduit for the disease to disseminate throughout Europe. However, the cadaver-catapulting probably didn't play a major part in spreading the pandemic – *Yersinia pestis* was a consummate killer even without human help.

One happy thought for the day is that sometimes modern-day terrorists are as inept as they are deranged. This was the case in the summer of 1993 when members of the doomsday cult/terrorist group Aum Shinrikyo ascended to the roof of their headquarters in Kameido, near Tokyo. They carried with them a liquid suspension of *Bacillus anthracis*, the anthrax bacteria. The assembled maladroit masterminds sprayed the suspension high into the air with the aim of raining down death on the unsuspecting citizens below. Thankfully they failed spectacularly, as it turned out the strain of anthrax they thought was a bioweapon was actually commonly used as a veterinary vaccine to protect animals from getting anthrax. A total of zero people were killed or injured that day.

While that summer's day consequently went by largely unnoticed, two years later the same group made their presence known by releasing sarin gas at five locations on the world's busiest subway system in the buzzing centre of Tokyo. That attack killed 13 people, and left another 5,500 with injuries that ranged from the mild to the severe. Had the cult used a deadly strain of weaponised anthrax for the underground attack the death toll would likely have been much, much higher.

In its inhaled form, anthrax is one of the deadliest diseases on the planet, killing 85–90 per cent of those infected. Central to its success is a potent three-part toxin made up of protective antigen (PA), lethal factor (LF) and oedema factor (EF). Together they form one of the most deadly biological weapons on the planet, yet individually each element is entirely harmless. It's only relatively recently that lab tests have demonstrated how this little triad of toxins are very much dependent on each other. PA is the muscled henchman of the group that attaches to the surface of our macrophages and gathers EF and LF to its side. It then slices a slit in the wall of the cell, like a burglar cutting a hole in a glass jewellery display case, which allows the other two toxins to penetrate deep inside. Through dark arts that are keeping many a studying scientist busy, EF and LF then cause the macrophages to become massively swollen and die. This approach slaughters these foot soldiers of our immune system and without macrophages to protect us the anthrax is free to multiply by the million, largely unchallenged. Fever, chills, shortness of breath and drenching sweats ensue, with death following swiftly for most.

For those with the inclination, making a weaponised version of this already deadly bacterium can be a surprisingly simple affair, achieved with fairly unexciting equipment costing a few thousand dollars. Near pure anthrax can be generated with a simple centrifuge that spins down the powder and separates out the spores. The next job is to get it really small because the perfect weaponised anthrax spore is less than a tenth of the thickness of a single human hair. This size allows it to travel quickly and easily along the maze of branching airways in the lungs. These branches become progressively smaller as they get deeper and end in the alveoli, which are tiny cloud-shaped air-sacs where we allow welcome guests like oxygen into our blood. It's also the entry point for anthrax, which lurks in the alveoli awaiting its cellular carriage: our macrophages. The macrophages are doing what they ought – sweeping up the dirt and debris sucked into the lungs every time we breathe – but unfortunately this means they take up the anthrax spores. Completely unaware that they have picked up a time bomb, the macrophages carry the spores inside them to the lymph nodes in the chest, which become the staging post for the dissemination of the disease. When they're good and ready the anthrax spores germinate and the bacteria burst out of the macrophage and spread through the bloodstream, causing an almost certainly fatal infection. This is in part because we have limited treatment options for taking on anthrax – all we have are antibiotics and intensive care units. The latter allows doctors to support or even temporarily replace the function of failing organs, hopefully long enough for the antibiotics and the immune system to see off the anthrax.

Although our treatments for anthrax and other killer bugs like Ebola are relatively limited, the twenty-first-century pharmacy is like a war chest for taking on diseases from cancer to carbuncles. There are remarkable feats of immunological alchemy being acted out every day at the frontiers of drug discovery and it is entirely feasible that one day there will be something to end anthrax's 90 per cent mortality rate. In our final chapter we will sneak a future-gazing peek at the cutting edge of drug discovery and see how it might save us from marching into the antibiotic apocalypse.

Clever Drugs: Immunological Alchemy

The members of the murderous cast of immune cells and messengers who populate the stories in this book have been inspiring scientists to forge new weapons and create their own lab-grown copycat killers. These scientists don't strive to simply observe and understand our body's defences, they hope to become the Sorcerer's Apprentice, to take the reins of this awe-inspiring army and harness its power to do their bidding. Some have succeeded. In this chapter we'll explore some of the most important immune–altering drug discoveries, bend our minds around a selection of the ethical quandaries they create and sneak a peek at the exciting pharmacopeia of the future being written by the immunological alchemists of today.

Copycats

As we've seen, the immune system is like a very complex game of Mouse Trap, except all the different interconnected pieces work together to catch invading micro-organisms instead of cheese-thieving mice. Changing the amount of any one of the cells or chemical messengers in the chain can have a significant impact on how well the system works as a whole. Therefore making man-made copies of these messengers is a common strategy for scientists trying to alter the immune response.

One example of this approach is the creation of synthetic interferons. As we saw in Chapter 3, interferons are a family of signalling molecules used by the immune system to switch on lots of different anti-intruder genes in our cells. The actions of these genes result in a less favourable environment for bacteria and viruses, effectively 'interfering' with their ability to multiply inside us. We've known about the microbial-thwarting talents of interferons for over 50 years, but for decades the scientists working on them couldn't find a way to apply these impressive little molecules to treating diseases. This was largely because it proved incredibly difficult to create pure enough interferons in large enough quantities to conduct high quality clinical trials – it was a bit like designing an experiment to assess the effects of alcohol intoxication with only a half-eaten box of chocolate liqueurs left over from last Christmas to test your theory. Basically it was never going to work. So despite interferons being an exceedingly effective strategy when deployed by the immune system, scientists and doctors just couldn't find a way to brandish them. As one poetic researcher put it, interferons were consequently cast into 'a scientific Siberia of sorts'. This exile lasted until the advent of gene-cloning techniques, when Interferon Alpha became one of the very first human genes to be cloned. Cloning enabled large-scale production and in 1991 Interferon Alpha became an approved treatment for chronic hepatitis C, a viral infection with few other effective treatment options. Today different types of interferon are used to combat various diseases including hepatitis B, multiple sclerosis and several forms of cancer.

Haute couture immunotherapy

While interferon copies are a useful tool, in some ways they are a relatively simple one because they can only be used to mimic the action of the original messenger. The next step up in terms of complexity in drug development is to create a designer version of one of the immune system's killer pieces, crafting tailored alterations with a specific task in mind. The classic example of haute couture immunotherapy is the monoclonal antibody (mAb). They are effectively antibodies that act like heat-seeking missiles designed to lock on to a single specific molecule. The brilliance of the mAb is that you can design one to lock on to just about any tiny molecule you desire. This means that if you want to deliver drugs or even radiation to a particular cell type, all you need to do is design a mAb that's shaped in such a way that it will bind to a unique marker on that cell. You don't need to know where all the cells are because your mAbs will do the legwork for you, incessantly scouring the body for their target destination like tiny, demented postal workers without a good union.

The applications of mAbs are legion: from scientists using fluorescent mAbs to illuminate the cells they study down a microscope, to doctors using therapeutic mAbs to systematically destroy the inaccessible innards of cancerous tumours. The earliest prototype mAbs were produced by Georges J. F. Köhler and César Milstein in 1975, and around 50 mAbs have since been approved across Europe and the US as treatments for a host of different diseases, with about four new mAbs joining the market every year. It's been estimated that by 2020 worldwide sales of mAbs will be worth in excess of $125 billion (£95 billion).

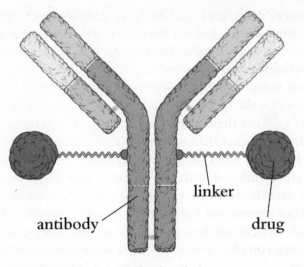

Figure 16.1 Drug delivery monoclonal antibody.

A vast constellation of mAbs can be made because they can vary in two ways: in the molecule they lock on to and in what they deliver. On the delivery end, some mAbs are essentially naked, and work by painting a bullseye on the surface of cells they attach to, labelling them for destruction by white blood cells in exactly the same way as any naturally created antibody. An example of a naked mAb is rituximab, which is designed to lock on to a marker called CD20 that's found in high levels on the surface of some B cells. Its penchant for CD20 means rituximab can be used to decimate diseases that produce an excess of B cells, such as non-Hodgkin's lymphoma (a type of cancer), by flagging the cells up to the immune system. Rituximab is now such a core part of the cancer doctor's medical bag that it's on the World Health Organisation's list of essential medicines for priority diseases.

Other mAbs can be modified to deliver chemotherapy agents directly to tumour cells. For instance, Mylotarg is a mAb designed to stick to a marker commonly found on leukaemia cells and which carries the sting of a powerful anticancer drug in its tail.* Kadcyla is another mAb with a chemotherapy tail, but it differs from Mylotarg in that it has a double-pronged attack. The target on the cancer cells that Kadcyla locks on to is called HER2 (human epidermal growth factor receptor 2), a growth receptor found in high numbers on the surface of some kinds of breast cancer. The cancer cells use their super-HER2 status to help them grow and divide quicker than normal cells, but if Kadcyla binds to the HER2 it blocks the growth signals getting through to the cell. Combined with the chemotherapy agent attached to its tail, Kadcyla packs a powerful one-two punch of blocking growth and killing cells. Impressive as the science of Kadcyla is, other elements of its bench-to-bedside journey are more complex, more murky; a subject we will return to later.

While cancer treatment is one of the main beneficiaries of the mAb revolution, it's not the only disease that can be treated with them. Infliximab uses the blockade technique of Kadcyla without the chemotherapy sting. Its target is an immune messenger called TNF-alpha (tumour necrosis factor alpha), which plays a big part in stirring up inflammation in various autoimmune diseases, including rheumatoid arthritis and Crohn's disease (which involves inflammation in the gut).

---

* Mylotarg is a bit of a complex cautionary tale about new medications. It was initially given fast-track approval by the FDA but on the condition that further trials were conducted after it was made available. The result of one such trial was that Mylotarg was voluntarily withdrawn from the US market because it seemed to offer no clinical benefit compared to existing treatments and had a higher mortality rate. This study conflicted with other evidence, which is why the drug is still available in Japan where regulators disagreed with the FDA's assessment of its risk-to-benefit profile. By their very nature, drugs at the cutting edge aren't as tried and tested as those that have been on the market for a long time. This is why it's incredibly important to monitor the impact of the drug once it's out on the market and being prescribed in the wild, rather than just in tightly controlled trials.

Although blocking TNF-alpha has a positive impact on these diseases, shutting down a major communication route in the immune system's chattering network can have other unintended consequences. It's a bit like Facebook – we all have friends who share irritating things, but if we cut off Facebook altogether we lose not only the mundane minutiae of people's dietary habits, but also Facebook's benefits. With infliximab, the reduction in TNF-alpha's immune chatter can increase the risk of some serious side-effects including pneumonia or septicaemia, and in children lymphoma, leukaemia and other cancers. The majority of people who use infliximab will not experience such serious negative effects, but it's a stark reminder of how big the consequences of manipulating even just one molecule in the immune system can be.

All of the mAbs we've explored so far are impressive, but riding at the very front of the exciting drug wave are the checkpoint inhibitor mAbs, including pembrolizumab, which help the immune system attack tumours. Tumour cells thrive in part by hiding from the immune system, sometimes using safety mechanisms built into the immune system to stop it attacking our own cells. One of the tumour techniques is to give any enquiring T cells a 'these aren't the cells you're looking for' handshake that sends them on their way in a deactivated state, unaware they have let the cancer cells off the hook. Checkpoint inhibitor mAbs bind to the T cell and prevent the deactivating handshake from happening. This leaves the T cell alert and able to recognise and destroy the cancer cells. Doctors monitoring a tumour after treatment with a checkpoint inhibitor see it seem to swell initially, but the increase in size is actually just T cells rushing into the tumour to attack it. Once the T cells get really stuck into killing the cancer cells, the tumour shrinks or disappears entirely. Pembrolizumab has been used to great effect to treat melanoma, a skin cancer we met in Chapter 14, which typically has a poor prognosis in the advanced stages. Prior to 2011, over 50 per cent of people with advanced melanoma died within a year, but in a trial of 644 people given pembrolizumab, 40 per cent were alive after three years and 15 per cent actually had no signs of cancer at all. These are game-changing statistics. However, harnessing the immune system to destroy tumours is no mean feat, and as with infliximab, the use of checkpoint inhibitors can come with serious side-effects. In 10 per cent of people, taking the brakes off the immune system causes an autoimmune reaction, with symptoms potentially affecting almost every system in the body, from diarrhoea to a rash to difficulty breathing. Thus anyone prescribing checkpoint inhibitors has to balance the risks against the benefits.

The cost of haute couture

Alas, haute couture immunotherapy, like its clothing counterpart, does not come cheap, and the ethics of the cost of immunotherapy are at least as complex as their science. To highlight this I'd like to tell you the tale of alemtuzumab. This is a plain Jane of a mAb, another example of a naked mAb that works just like a regular antibody but in this case has been tailored to attach to CD52, a marker displayed in high numbers on the surface of certain T and B cells. This makes it a useful treatment for some leukaemia patients who don't respond to standard treatments. However, a few years after its release as a leukaemia drug under the brand name Campath, alemtuzumab was shown to have even more impressive benefits for patients with multiple sclerosis (MS). A clinical trial published in the *Lancet* in 2012 found that 51 per cent of patients treated with interferon beta 1a (the standard treatment) relapsed, compared with just 35 per cent of patients in the alemtuzumab group. While the mechanism underlying this effect wasn't clear, the promising results led doctors to prescribe the drug 'off-label' for MS.[*] Having realised that alemtuzumab's role in the leukaemia market was declining and seeing the massive potential for it to become

---

[*] 'Off-label' prescribing involves prescribing a drug for a purpose or in a dose that's different from the one it has regulatory approval for.

the best drug for MS on the market, its makers took a very logical step: they stopped selling the drug in the US and Europe. Not because it wasn't safe or it didn't work, but because they wanted to re-enter it into the market rebranded, and more importantly re-priced, as a treatment for multiple sclerosis, an area with limited treatment options.* So Lemtrada (aka the drug formerly known as Campath) was born.

Under the guise of Campath, alemtuzumab had cost around $60,000 (£46,000) when administered three times a week over a 12-week course. However with its new 2014 Lemtrada facelift, half the dose for five days, followed by a further three-day course 12 months later, cost about $158,000 (£120,000). The cost discrepancy between the two caused understandable uproar among some in the MS community. Nevertheless, the National Institute for Health and Care Excellence (NICE), which is the UK body that evaluates the cost-effectiveness of drugs, still approved Lemtrada's use in the NHS. Ultimately when compared to the cost of other MS drugs, its superior results meant it still made good financial sense.

Yet the general idea of a drug increasing in price when it's proven to work well for a disease probably raises all but the most cynical of eyebrows. A particularly extreme example of this hit the headlines in late 2015 when Martin Shkreli, the CEO of Turing Pharmaceuticals, announced that having acquired the rights to a drug used by AIDS patients, his company would be increasing the price tag by over 5,000 per cent. The drug in question was Daraprim, a treatment for the parasitic infection toxoplasmosis, which tends to affect immunocompromised people including those with AIDS. It cost about a dollar to manufacture each Daraprim tablet and

---

* While you might think withdrawal of drugs from the market is primarily a safety issue, this isn't the case; in the US from 1980 to 2009, 118 drugs were withdrawn from the market, but only 22 per cent of these were withdrawn for safety reasons. In the UK, of all new drugs approved from 1972 to 1994, 59 per cent of the subsequent withdrawals were for commercial reasons.

the drug had been retailing in the US at about $13.50 (£10.28) per dose until Turing Pharmaceuticals acquired it and bumped the price to $750 (£571) per dose. Shkreli pointed to the fact the $1 (76p) production cost didn't include key expenses like marketing or distribution, and argued that the profit would be reinvested in new drug development. Unsurprisingly this didn't stop him becoming the poster child of corporate greed,[*] and even an industry insider described him as 'like a male version of Cruella de Vil'. In this case there was a happy ending because the patent situation meant another company could offer a drug that was a combination of Daraprim's active ingredient plus another medication for as little as $100 (£76) for 99 tablets.

However, there are plenty of drugs with patents in full force that prevent any other company from offering a cheaper alternative, a fact that can lead to astronomical prices. Many argue this reality simply reflects the nature of developing cutting-edge drugs like mAbs and the other therapies in this chapter. Pharmaceutical companies take on big risks when they develop new drugs and therefore want a guarantee of a big return if the gamble pays off. In 2015, a report by the US Tufts Center for the Study of Drug Development calculated the cost to a pharmaceutical company of getting a new drug successfully to market to be $2.6 billion (£1.98 billion). Charities and patient groups have disputed this figure, and some academics have pointed to the significant financial support the Tufts Center itself receives from drug companies. Yet whether the number is $2.6 billion or Deloitte's more conservative estimate of $1.4 billion (£1.08 billion), the point remains that this is a massive financial commitment.

---

[*] His less than sympathetic image wasn't aided by revelations of regulatory and criminal investigations at hedge funds he once managed. Or the claim he was expelled from a previous pharmaceutical company he ran 'because of serious concerns about his conduct'. Or his discontent that a donation he sent to presidential hopeful Bernie Sanders was rejected and instead handed to a health clinic specialising in HIV/AIDS.

A sizeable chunk of the Tufts figure comes from the spectacularly high failure rate of new drugs. For instance, Pfizer dropped $800 million (£619 million) on torcetrapib, a new drug for high cholesterol that bombed spectacularly. This isn't uncommon and the reality is that most new compounds will never see the pharmacy counter. Those that do will have taken on average 12.5 years to make the perilous journey from promising idea to prescription pad. In the light of this you can understand why companies are keen to maximise their profit margins when a drug does make it to market.

Yet this creates an explosion of ethical quandaries, especially with costly cancer drugs, many of which are immunotherapies. One of the biggest questions is: in cold, hard cash terms, exactly how much is a month of extra life for a terminally ill person worth? When it was approved by the FDA in 2011, the checkpoint inhibitor mAb ipilimumab had been shown to extend the life of skin cancer patients with metastatic melanoma by 3.7 months over standard treatment in previously treated patients and 2.1 months in previously untreated patients. The cost was $120,000 (£91,000) for four doses, which cancer expert Dr Leonard Saltz from Memorial Sloan Kettering Cancer Center noted was gram-for-gram 4,000 times the price of gold. He also calculated that the cost for a patient at the highest dose of pembrolizumab, which we encountered earlier in the chapter, could be reasonably estimated at $1,009,944 (£768,837) per year. If they spent every cent of their take-home pay on pembrolizumab, it would take someone on the 2015 federal minimum wage 72 years and 6 months to pay for a year's supply of the drug.

Of course insurance companies and government healthcare systems have considerably more spending power than individuals, but they still have to decide what to fund. In a world where the state has infinitely deep pockets we would probably all agree that every patient should be given access to the medications they need, especially when the benefits are life-extending. However with the prices of some drugs so high, there isn't a country on the planet with pockets deep enough to pay for everyone to get access to every cutting-edge treatment, which means choices have to be made. In the UK, NICE is the organisation of experts responsible for making these difficult choices and deciding which drugs should be available through the National Health Service. They receive little thanks for this in the media, generating headlines like: 'Betrayal of 20,000 cancer patients: Rationing body rejects ten drugs (allowed in Europe) that could have extended lives'. In a move since described by the CEO of a cancer charity as 'a "quick fix" to politically damaging newspaper headlines about cancer patients dying because their "life-saving" treatments had been turned down by NICE', Prime Minister David Cameron announced the Cancer Drugs Fund. Launched in 2011, its purpose was to provide funding for drugs NICE had yet to evaluate or had turned down on the grounds that they weren't cost effective or didn't work well enough. While there is no question patients benefited from the fund at one of the hardest times imaginable in their lives, and it was applauded by several patient groups and some segments of the media, its existence raised some difficult questions. For instance, why should cancer be prioritised over other equally devastating, life-limiting diseases? What makes uncontrolled cell division so much more worthy of that funding?

As a thought experiment, if I gave you the purse strings to the £1 billion ($1.3 billion) of public money dedicated to the Cancer Drugs Fund and asked you to spend it on health, what criteria would you use to decide how to spend it fairly? Would you make a list of the 10 most common diseases? Or the 10 most deadly diseases? Would you prioritise certain groups of people? Would you consider the extent to which a person's behaviour caused the disease? Would you consider what taxpayers think? A survey of over 4,000 British adults found people supported the use of criteria such as disease severity, having wider societal benefits and innovative treatments (as long as they offered substantial health benefits). They weren't supportive of the idea of the Cancer Drugs Fund.

If, like me, you're an evidence-based medicine nerd, you might want to fund the most effective interventions, those with the biggest bang for your billion bucks. Sounds like a plan, but even if there is evidence available,* how do you compare the benefits of a dementia drug with a programme to encourage safe sex among teenagers? NICE uses a system based on calculating the number of Quality Adjusted Life Years (QALYs) gained by an intervention, be it a drug or something else like counselling or an exercise programme. A QALY is a 'measure of the state of health of a person or group in which the benefits, in terms of length of life, are adjusted to reflect the quality of life. One QALY is equal to one year of life in perfect health.' For instance, NICE estimates that a brief intervention plus self-help for smoking-cessation costs the NHS £372 ($489) per QALY, while an exercise prescription to increase physical activity costs £77 ($101) per QALY. This compares to an estimated £166,000 ($218,000) per QALY for Kadcyla, a mAb used to treat breast cancer that NICE decided was too expensive to be funded on the NHS. The missing narrative behind the damning headlines is that there is a finite pot of money, and every pound spent on Kadcyla is a pound not spent on other, more cost-effective treatments. Of course there is more complexity to spending money than cost-effectiveness. In the case of cancer treatments, it can be asking if you would deny Katie, loving mother and wife, a few precious months of extra life to say all the things she thought she had for ever to say to her kids and husband. Ethicists call this the Rule of Rescue – the need to rescue identifiable individuals. We can see very visibly the individuals being denied the cancer drugs, and the very immediate, heart-breaking impact on them and their loved ones. Less visible are the people who will die from smoking-related illnesses in years to come and could have been saved by smoking-cessation. Or the people who die from heart disease that could have been prevented, had they been helped to be more physically active. Yet just because we can't see

---

* This is a big if. High-quality evidence applicable to your particular circumstances is like finding the pot of gold at the end of rainbow *and* the leprechaun who owns it.

them so easily doesn't make their lives any less valuable, or their deaths any less tragic.

Moreover, we are far more likely to have a friend or loved one who could have benefited from these interventions than from a super-costly cancer drug. According to researchers at the University of York, for every QALY gained through the Cancer Drugs Fund, five QALYs were lost across the NHS as a whole. The prime minister set up the Cancer Drugs Fund because nobody should be denied treatment, but sadly that's not in his power to achieve, because the NHS does not have unlimited funds. Some have argued that the biggest beneficiaries of the Cancer Drugs Fund have been the pharmaceutical companies, whose drugs were purchased by the NHS. However we'll never know because no data was gathered to assess the fund's impact. Ultimately even the Cancer Drugs Fund wasn't bottomless and a £100 million ($131 million) overspend was anticipated for 2014/15. The resource allocation problem isn't going anywhere, and as more of us live longer and more mAbs and other expensive drugs hit the market, the question remains: how should you spend your billion bucks?

Avoiding the antibiotic apocalypse

What do you expect to die of? Statistically speaking, a sensible guess in much of the Western world would probably be a chronic disease like diabetes, cancer or heart disease. In the US, 9.3 per cent of Americans have diabetes; one person dies of heart disease every three minutes in the UK; and half of us will develop cancer at some point in our lives. The Great Pox (aka syphilis), scarlet fever and gangrene probably didn't enter your imagination. Thanks to antibiotics many of the once fear-inspiring infections have long since lost their edge. Yet some people think that we are standing on the precipice of a human disaster and we'll shortly fall over the edge and find ourselves plunging headlong into the medical dark ages of a post-antibiotic era – a time when even common bacteria are resistant to attack by antibiotics. In such a world even minor surgery would no longer be safe, and common operations like caesarean sections would have to stop. Even a simple cut could be transformed into unstoppable blood-poisoning. Equally, life-saving treatments like chemotherapy or organ transplants would become lethal because doctors would no longer be able to safely suppress the immune system, lacking the safety net of antibiotics for infections. While it would be a disaster for us, it would be an opportunity of epic proportions for the infectious diseases of old to reassert themselves. The seeds of this dystopian scene of an antibiotic apocalypse are being sown today through a combination of how we use our existing antibiotics and the desperately dry pipeline of new antibiotics.

There is a multitude of ways bacteria can become resistant to an antibiotic, all of which begin with a mutation in a bacterium that gives it a survival advantage when dunked in the drugs. For instance, they can change their surface so the antibiotic can no longer get inside the bacterium, or develop antibiotic bilge pumps that prevent the drug from building up. Alternatively they may create new metabolic pathways that mean they can bypass the antibiotic's roadblock. One of the common resistance mechanisms employed to circumvent penicillin is the production of a type of enzyme called a beta-lactamase, which breaks up a critical ring structure within the antibiotic. All of these antibiotic-avoiding techniques (and more) are the result of mutations in bacterial genes that give their host a survival advantage when exposed to antibiotics. The problem is we are setting up easy antibiotic assault courses all over the place, by not finishing a course of antibiotics, or taking them unnecessarily, providing a perfect training ground for the bacteria to become resistant by exposing them to antibiotics again and again until some of them adapt instead of dying.

Health professionals can also foster resistance by prescribing antibiotics that provide broad protection rather than tailoring their prescription to the microbe they're trying to kill, or by failing to prescribe the right drug in the right dose. In December 2015, prescribing the wrong mix of drugs was blamed for the creation of the terrifyingly titled super gonorrhoea in Leeds. Gonorrhoea is a sexually transmitted infection that can ordinarily be knocked out with a combination of two antibiotics, but there were concerns that some patients were only receiving one of the drugs, allowing the infection to fester and buying the bacteria more time to develop resistance. The problem was so significant that England's chief medical officer wrote to all GPs and pharmacies in England to warn them that 'Gonorrhoea is at risk of becoming an untreatable disease due to the continuing emergence of antimicrobial resistance.' With a colossal 19 per cent increase in gonorrhoea cases in England from 2013 to 2014, the prospect of this infertility-causing infection becoming invincible is a public health disaster in the making. Unsurprisingly, with examples of resistance like this, there are campaigns to discourage patients from expecting an antibiotic solution to their ills, and doctors' prescribing is subject to increasing scrutiny. Yet people are only one piece of the puzzle, as these training camps are also being set up on an industrial scale in farm animals. In some countries over 50 per cent by weight of antibiotics used across the nation are being dished out in farmyards rather than pharmacies. This is because antibiotic use allows farmers to rear animals in more tightly confined (and therefore profitable) spaces. In some countries vets can also have a significant financial incentive for prescribing them, with at least 40 per cent of their income coming from drug sales.

In the same month that super gonorrhoea was surging in Leeds, the dangers of the agricultural use of antibiotics were highlighted by forbidding headlines such as 'Bacteria that resist "last antibiotic" found in UK'. The antibiotic in question was colistin, a drug we've had for decades but that is held back as a treatment of last resort due to its toxic effects on the kidneys. Reports of resistance had already been issued from other countries including China, Denmark, Thailand and France, prompting the UK to test 24,000 historic samples of bacteria collected from infections in people from 2012 to 2015, as well as testing farms for the resistant bacteria. They found colistin-resistant bacteria in 15 of the 24,000 samples and on three pig farms. Thankfully the bacteria in question were found to be susceptible to other antibiotics, a fact that prevented them from being the microbial horsemen of the apocalypse. However, despite the tiny proportion of colistin-resistant organisms, this is a cause for concern because bacteria are genetically promiscuous and can pass resistance genes between species. They do this by replicating circles of DNA and then passing one copy down a long thin tube called a pilus, which they poke into the side of the other bacteria. It's a sort of bacterial sex, except the bacterial pili are very fragile, prone to breaking off, and need to be frequently replaced. There is no bacterial contraceptive to stop the spread of resistance between species once it has developed, so the pressure is on to stop bacteria ever developing resistance to colistin. Despite being rarely used in people, 837kg (1,845lb) of colistin was reportedly used by British farms in 2014. This fact fuelled calls to cut the use of the antibiotic in animals to eliminate a major colistin-resistance training ground. While this is a step in the right direction, antibiotic resistance is a global issue because bacteria don't need passports. So with China using 12,000 tonnes of colistin in animals annually, the rise of colistin-resistance is currently an inevitability.

In truth it's impossible to prevent resistance for ever – we're ultimately fighting evolution – so all we can hope to do is slow the progression. This makes the creation of new antibiotics absolutely vital if we're to avoid the antibiotic apocalypse. As the Director-General of the World Health Organisation has said, 'The trends are clear and ominous. No action today means no cure tomorrow.' In a lab in Singapore, synthetic biologist Matthew Chang is taking action. He is a man with a vision: a microscopic army, capable of seeking out and destroying bacteria in the body. In version 1.0 of this dream, he took *E. coli* and designed them to be kamikaze killers. His suicidal minions were engineered to seek out a pneumonia-causing bacterium by the name of *Pseudomonas aeruginosa* and explode in its presence, releasing a toxic cloud of a *Pseudomonas*-slaughtering chemical called pyocin. Version

2.0 was more terminator than kamikaze because it could secrete its killer payload, a chemical called microcin S, instead of having to explode. One of the things that makes Chang's bacteria an exciting prospect for replacing antibiotic therapy is the fact that unlike antibiotics, which tend to be somewhat non-selective and kill both our 'good' and 'bad' bacteria, the terminators are targeted killers. They can do this because *E. coli* seek out messenger beacons, which are thrown out into the ether by the *Pseudomonas* bacteria as they try to get a feel for the size of their colony. The *E. coli* detect and move towards these beacons, which also stimulate the little terminators to produce and release their chemical warfare. So far these weaponised *E. coli* have only been given to mice, but the results have been promising, with the numbers of *Pseudomonas* in the mouse poo reduced with no detectable ill effects. There are a lot of studies that separate us from a future where these engineered bacteria are widely used, but perhaps one day, just as we currently consume pots of 'good' bacteria, we will also swallow little soldiers.

In fact this seemingly future-gazing idea is positively old hat to people in Russia, Poland and Georgia, where it's long been commonplace to swallow medicinal viruses called bacteriophage. These naturally occurring viruses infect and kill bacteria, and for almost a century people have been using them to help the immune system fight off bacterial infections. They're like microbial snowflakes, with no two identical – a fact that reduces the chances of bacteria developing resistance to them. They've been widely used in Russia, prepared as everything from creams to enemas, injections to tampons, to fight a range of bacterial infections. However little of the Russian experience has escaped into the English medical literature, in part because of the language barrier and in part because the results of scientific enquiry funded by the military, such as phages for treating infections acquired in the battlefield, automatically became state secrets.

While enthusiasm for the phage isn't shared in present-day western Europe, the earliest known use of human bacteriophage therapy comes from France in 1919. Intrepid microbiologist Félix d'Hérelle cultivated bacteriophage from the stools of soldiers and fed them to children suffering from severe dysentery in the Hôpital des Enfants Malades in Paris. The children recovered from dysentery, but rather than do more clinical trials, d'Hérelle set about a series of exacting experiments to classify different kinds of phage, writing several tomes on the subject. Others did continue with clinical research, using phages to treat a variety of maladies from cholera to boils. Eventually d'Hérelle was ready to release the knowledge he'd gained, and launched the world's first commercial phage therapies, including Bacté-Intesti-Phage, Bacté-Dysentérie-Phage, Bacté-Pyo-Phage and Bacté-Rhino-Phage. Surgeons used them to reduce mortality after typhoid-induced gut perforations, following reports that squirting phage solution into the abdominal cavity after surgery seemed to drastically cut mortality from 85 per cent to 20–35 per cent. There were even reputable European laboratories, including the esteemed Pasteur Institute of Lyon, making their own phage cocktails that doctors could use. In the five years from 1969 to 1974 the Pasteur Institute identified and purified 476 different phages, many of which targeted common Staphylococcal infections. They used them to treat all sorts of Staph infections that had resisted antibiotics, from pus growing on heart valves to chronic infections boring through bones.

This continued until as recently as the mid-1990s, however, France was something of an exception in Western medicine as phage therapy was largely rebuffed elsewhere in Europe and in the United States. One review of phage therapy suggested this discrepancy could be explained by the idea that the French excelled at farming phages, thanks in part to the painstaking work of d'Hérelle, while phages being prepared elsewhere had quality control issues. Certainly one investigation of commercial phage therapies from three well-known US pharmaceutical companies in the 1930s found they all had problems with quality control and stability. However the real body-blow to phage therapy in Western medicine in the 1930s and 1940s came from two unfavourable reviews of phage studies published in the *Journal of the American Medical Association*. These were limited to studies published in the English language, which given that much of the phage work was done in France may have gone some way to explaining their findings. Moreover most of the studies of phage therapy at that point were also small and of questionable quality, making it difficult to provide definitive evidence of whether phages worked.

This was still largely true until very recently, when the inexorable rise of antibiotic resistance led to renewed interest in phages and prompted the funding of Phagoburn, which was billed as the first international clinical study on phages meeting international standards in clinical evaluation. The trial involved 220 patients across 11 centres in France, Belgium and Switzerland and began in July 2015, receiving €3.85 million ($4.3 million, £3.28 million) in EU funding to evaluate the effectiveness of phage therapy in treating antibiotic-resistant infections in burn victims. The phage cocktail was provided by a company that has scoured a variety of sources, from sewage to river water, to identify viruses capable of infecting and killing bacteria that cause human disease; so far they have a repertoire of over 1,000 bacteriophages. The trial's final results are yet to be published at the time of writing, but they have the potential to either build or demolish one of the greatest hopes we have of fighting antibiotic resistance.

Yet even if phage therapy triumphs in a multitude of high-quality clinical trials, there remains a barrier at least as challenging as whether it works or not: whether it is profitable. The fact that phages have been in use for over a century makes it nigh-on impossible for a pharmaceutical company to assert an intellectual property claim on them. Any bid in the US to patent a phage found in nature is unlikely to be successful following a 2013 Supreme Court ruling that barred patents on natural genes. However, the possibility is still open of patenting a genetically engineered phage, something that will pique interest in the work of Timothy Lu. Lu is a synthetic biologist from the Massachusetts Institute of Technology whose group has produced a phage that kills *E. coli* with specific antibiotic-resistance genes, while leaving other *E. coli* in peace. It's shown promising results in worms, but there's a great expanse of experimental traps, practical hurdles and an ethical minefield between improved worm health and human trials.

Yet we must persevere. By 2050 the problem of antibiotic-resistant infections is predicted to cause at least 10 million extra deaths a year, and between now and then the cost to the global economy is predicted to be $100 trillion (£80 trillion). From 1940 to 1962, more than 20 new classes of antibiotics hit the market, but since 1962 just two more have appeared. The most recent new class of antibiotic, the lipopeptides, were a discovery of the late 1980s – a time so long ago, shoulder pads were cool and New Kids On The Block dominated the charts. While there have been new antibiotics approved by the FDA since then, they've been within existing classes and therefore work in similar ways and have similar vulnerabilities to resistance as other widely used antibiotics. However, the answer to our woes may be right beneath our feet. Literally. In January 2015, a group of scientists caused a stir with a *Nature* paper entitled 'A new antibiotic kills pathogens without detectable resistance'. They described their discovery of a bacterium in soil that produced an antibiotic chemical they named teixobactin. What made this discovery exciting was that teixobactin seemed to have a totally new way of stopping bacteria from making their cell wall, and its target was common to lots of bacteria. This explained why none of the various mutant strains of bacteria, including the one that causes TB, were resistant to teixobactin. There are many, many years of testing before teixobactin hits the market, and not all bacteria have the vulnerability of a cell wall, but the possibility of a new method of attack is an exciting and much-needed glimmer of light in the antibiotic apocalypse tunnel. Just as exciting is the idea that soil could be a much-needed new pipeline for antibiotic discovery, and it's not alone.

Another new avenue for investigation is almost as glamorous as soil: snot. In 2016 the wonderfully dubbed 'snot wars' study was published in *Nature* and reported the discovery of lugdunin – an antibiotic secreted by bacteria in our noses capable of taking on superbugs like MRSA. Though it may not seem like it, our nostrils are prime real estate and rival bacteria fight each other for resources, a fight which includes chemical warfare. The researchers had found that people whose noses contained *Staphylococcus lugdunensis* were less likely to have another bacterium called *Staphylococcus aureus*. They attributed this to an antibiotic chemical being secreted by *S. lugdunensis*, which they named lugdunin. As with teixobactin, there is a long road before lugdunin makes it out of the lab and into our hospitals, but it points us at a previously untapped source of new antibiotics: the human body. Like so much of what we've explored in this book, this discovery is a beautiful reminder that of all the wonders of the world, the one with the most potential for changing the face of medicine and letting us cure the incurable is us.

References

This isn't an exhaustive list of all the sources of information I used when writing the book; instead it's a selection of the most significant ones. I've included links, but the internet is forever in flux so some of them may not work by the time the book hits the shelves and you try them ... in which case a little Googling is your best plan!

Chapter 1

Ahmed, N. 2005. 23 years of the discovery of Helicobacter pylori: Is the debate over? *Annals of Clinical Microbiology and Antimicrobials* 2005, 4: 17. Available from: www.ncbi.nlm. nih.gov/pmc/articles/PMC1283743/

Roberts, C. S. 1990. William Beaumont, the Man and the Opportunity. In: Walker, H. K., Hall, W. D., Hurst, J. W. (eds). *Clinical Methods: The History, Physical, and Laboratory Examinations*. 3rd edition. Boston, Butterworths.

University of Leeds 2003. Layers in the Epidermis. *The Histology Guide*. Available from: www.histology.leeds.ac.uk/skin/epidermis_layers.php

Chapter 2

Brinkmann, V. and Zychlinsky, A. 2012. Neutrophil extracellular traps: Is immunity the second function of chromatin? *Journal of Cell Biology* September 2012, 198 (5): 773–783. Available from: jcb.rupress.org/content/198/5/773.full

Perrone, L. A. et al. 2008. H5N1 and 1918 Pandemic Influenza Virus Infection Results in Early and Excessive Infiltration of Macrophages and Neutrophils in the Lungs of Mice. *PLoS Pathogens* 2008, 4(8): e1000115. Available from: journals. plos.org/plospathogens/article?id=10.1371/journal. ppat.1000115

Shooter, R. A. et al. 1980. *Report of the investigation into the cause of the 1978 Birmingham smallpox occurrence.* The House of Commons. Available from: www.gov.uk/government/uploads/system/uploads/attachment_data/file/228654/0668.pdf.pdf

Chapter 3

Ali, Y. M. et al. 2012. The Lectin Pathway of Complement Activation is a Critical Component of the Innate Immune Response to Pneumococcal Infection. *PLoS Pathogens* 2012, 8 (7): e1002793. Available from: www.plospathogens. org/article/info%3Adoi%2F10.1371%2Fjournal. ppat.1002793

Commins, S. P. et al. 2010. Immunologic messenger molecules: cytokines, interferons, and chemokines. *Journal of Allergy and Clinical Immunology* February 2010, 125 (2 Suppl. 2): S53–72. Available from: www.jacionline.org/article/S0091-6749(09) 01075-6/fulltext#sec11

Yiu, H. H. et al. 2012. Dynamics of a Cytokine Storm. *PLoS ONE* 2012, 7 (10): e45027. Available from: www.plosone.org/article/info%3Adoi%2F10.1371%2Fjournal.pone.0045027

Chapter 4

Brink, J. G. 2009. The first human heart transplant and further advances in cardiac transplantation at Groote Schuur Hospital and the University of Cape Town. *Cardiovascular Journal of Africa* February 2009, 20 (1): 31–35. Available from: www.ncbi.nlm. nih.gov/pmc/articles/PMC4200566/

Pearsall, P. 2002. Changes in Heart Transplant Recipients That Parallel the Personalities of Their Donors. *Journal of Near-Death Studies* 2002, 20 (3). Available from: www.newdualism. org/nde-papers/Pearsall/Pearsall-Journal%20of%20Near-Death%20Studies_2002-20-191-206.pdf

Raya-Rivera, A. M. et al. 2014. Tissue-engineered autologous vaginal organs in patients: a pilot cohort study. *Lancet* 26 July 2014 , Vol. 384, Issue 9940: 329–336. Available from: www. thelancet.com/pdfs/journals/lancet/PIIS0140-6736(14)60542-0.pdf

Snyder, T. M. et al. 2011. Universal noninvasive detection of solid organ transplant rejection. *Proceedings of the National Academy of Sciences USA* April 2011, 12 108 (15): 6229–6234. Available from: www. ncbi.nlm.nih.gov/pmc/articles/PMC3076856/

Chapter 5

Ardissone, A. N. 2014. Meconium Microbiome Analysis Identifies Bacteria Correlated with Premature Birth. *PLoS ONE* 9 (3): e90784. Available from: journals.plos.org/plosone/ article?id=10.1371/journal.pone.0090784

Fitz-Gibbon, S. et al. 2013. Propionibacterium acnes strain populations in the human skin microbiome associated with acne. *Journal of Investigative Dermatology* September 2013, 133 (9): 2152–2160. Available from: www.ncbi.nlm.nih.gov/ pmc/articles/PMC3745799/

Groeger, D. et al. 2013. Bifidobacterium infantis 35624 modulates host inflammatory processes beyond the gut. *Gut Microbes* 1 July 2013, 4 (4): 325–339. Available from: www.ncbi.nlm. nih.gov/pmc/articles/PMC3744517/

Hsiao, E. Y. et al. 2013. Microbiota Modulate Behavioral and Physiological Abnormalities Associated with Neurodevelopmental Disorders. *Cell* 19 December 2013, 155 (7): 1451–1463. Available from: www.cell.com/abstract/S0092-8674 (13)01473-6

Messaoudi, M. et al. 2011. Assessment of psychotropic-like properties of a probiotic formulation (Lactobacillus helveticus R0052 and Bifidobacterium longum R0175) in rats and human subjects. *British Journal of Nutrition* March 2011, 105 (5): 755–764. Available from: www.ncbi.nlm.nih.gov/pubmed/20974015

Morton, E. R. et al. 2015. Variation in Rural African Gut Microbiota Is Strongly Correlated with Colonization by Entamoeba and Subsistence. *PLoS Genetics* November 2015, 11 (11): e1005658. Available from: www.ncbi.nlm.nih.gov/ pmc/articles/PMC4664238/

Nood, E. van et al. 2013. Duodenal Infusion of Donor Feces for Recurrent Clostridium difficile. *New England Journal of Medicine* 2013, 368: 407–415. Available from: www.nejm.org/doi/ full/10.1056/NEJMoa1205037#t=articleBackground

OpenBiome available at: www.openbiome.org/

Ridaura,V. K. et al. 2013. Gut Microbiota from Twins Discordant for Obesity Modulate Metabolism in Mice. *Science* 6 September 2013, 341 (6150). Available from: www.sciencemag.org/content/341/6150/1241214.full

Ursula, L. K. et al. 2012. Defining the Human Microbiome. *Nutrition Reviews* August 2012, 70 (Suppl. 1): S38–S44. Available from: www.ncbi.nlm.nih.gov/pmc/articles/PMC3426293/

Chapter 6

Anon., 1923. Medicine: Voronoff and Steinach. *Time*, 30 July 1923, Vol. 1, Issue 22.

Bouman, A. et al. 2005. Sex hormones and the immune response in humans. *Human Reproduction Update* July/August 2005, 11 (4): 411–423. Available from: humupd.oxfordjournals.org/content/11/4/411.full

Broomfield, J. J. et al. 2014. Maternal tract factors contribute to paternal seminal fluid impact on metabolic phenotype in offspring. *Proceedings of the National Academy of Sciences* 2014, 111 (6). Available from: www.pnas.org/content/early/2014/01/23/1305609111.full.pdf

Fox, C. A. et al. 1973. Continuous measurement by radio-telemetry of vaginal pH during human coitus. *Journal of Reproduction and Fertility* April 1973, 33 (1): 69–75. Available from: www.reproduction-online.org/content/33/1/69.full.pdf+html

Kavanagh, J. et al. 2001. Sexual intercourse for cervical ripening and induction of labour. *Cochrane Database of Systematic Reviews* 2001, (2): CD003093. Available from: www.ncbi.nlm.nih.gov/pubmed/11406072

Robertson, S. A. et al. 2002. Transforming growth factor beta – a mediator of immune deviation in seminal plasma. *Journal of Reproductive Immunology* October–November 2002, 57 (1–2): 109–128. Available from: www.jrijournal.org/article/S0165-0378(02)00015-3/abstract

Chapter 7

Mor, G. and Cardenas, I. 2010. The Immune System in Pregnancy: A Unique Complexity. *American Journal of Reproductive Immunology* June 2010, 63 (6): 425–433. Available from: www.ncbi.nlm.nih.gov/pmc/articles/PMC3025805/

Chapter 8

Cairncross, S. et al. 2002. Dracunculiasis (Guinea Worm Disease) and the Eradication Initiative. *Clinical Microbiology Reviews* April 2002, 15 (2): 223–246. Available from: www.ncbi.nlm. nih.gov/pmc/articles/PMC118073/

Hansen, R. D. E. et al. 2011. A worm's best friend: recruitment of neutrophils by Wolbachia confounds eosinophil degranulation against the filarial nematode Onchocerca ochengi. *Proceedings of the Royal Society B: Biological Sciences* 7 August 2011, 278 (1716): 2293–2302. Available from: www.ncbi.nlm.nih.gov/ pmc/articles/PMC3119012/

Helmby, H. 2015. Human helminth therapy to treat inflammatory disorders – where do we stand? *BMC Immunology* 16: 12. Available from: www.ncbi.nlm.nih.gov/pmc/articles/ PMC4374592/

Janeway, C. A. et al. 2001. Pathogens have evolved various means of evading or subverting normal host defences. In: *Immunobiology: The Immune System in Health and Diseases*, 5th edition. New York, Garland Science. Available from: www. ncbi.nlm.nih.gov/books/NBK27176/

Wekerle, H. and Sun, D. 2010. Fragile privileges: autoimmunity in brain and eye. *Acta Pharmacologica Sinica* 31: 1141–1148. Available from: www.nature.com/aps/journal/v31/n9/full/ aps2010149a.html

World Health Organisation. Updated January 2017. *Trypanosomiasis, human African (sleeping sickness).* Available from: www.who. int/mediacentre/factsheets/fs259/en/

Chapter 9

Alberts, B. et al. 2002. The Generation of Antibody Diversity. In: *Molecular Biology of the Cell.* 4th edition. New York, Garland Science. Available from: www.ncbi.nlm.nih.gov/books/ NBK26860/

Maul, R.W. and Gearhart, P.J. 2010. AID and Somatic Hypermutation. *Advances in Immunology* 105: 159–191. Available from: www. ncbi.nlm.nih.gov/pmc/articles/PMC2954419/

National Cancer Institute 2013. *CAR T-Cell Therapy: Engineering Patients' Immune Cells to Treat Their Cancers.* Available from: www.cancer.gov/cancertopics/research-updates/2013/ CAR-T-Cells

Chapter 10

Deer, B. 2011. How the case against the MMR vaccine was fixed. *The BMJ* 342: c5347. Available from: www.bmj.com/content/342/bmj.c5347

Dove, A. 2005. Maurice Hilleman. *Nature Medicine* 11, S2. Available from: www.nature.com/nm/journal/v11/n4s/full/nm1223.html

Public Health England. *Immunisation against infectious disease.* Available from: www.gov.uk/government/collections/immunisation-against-infectious-disease-the-green-book#the-green-book

Riedel, S. 2005. Edward Jenner and the history of smallpox and vaccination. *Proceedings (Baylor University Medical Center)* January 2005, 18 (1): 21–25. Available from: www.ncbi.nlm.nih.gov/pmc/articles/PMC1200696/

Siegrist, C. A. *Vaccine immunology.* Available from: www.who.int/immunization/documents/Elsevier_Vaccine_immunology.pdf

Vaccine Confidence Project 2015. *The State of Vaccine Confidence 2015.* Available from: www.vaccineconfidence.org/The-State-of-Vaccine-Confidence-2015.pdf

Wakefield, A. J. et al. 1998. Ileal-lymphoid-nodular hyperplasia, non-specific colitis, and pervasive developmental disorder in children. *Lancet* 28 February 1998, Vol. 351, Issue 9103, 637–641. Retracted but available from: www.thelancet.com/journals/lancet/article/PIIS0140-6736(97)11096-0/abstract

Chapter 11

Anagnostou, K. et al. 2014. Assessing the efficacy of oral immunotherapy for the desensitisation of peanut allergy in children (STOP II): a phase 2 randomised controlled trial. *Lancet,* Vol. 383, Issue 9925, 1297–1304. Available from: www.thelancet.com/journals/lancet/article/PIIS0140-6736(13)62301-6/fulltext

Galli, S. J. et al. 2008. The development of allergic inflammation. *Nature,* Vol. 454, 445–454. Available from: www.nature.com/nature/journal/v454/n7203/full/nature07204.html

Graham-Rowe, D. 2011. Lifestyle: When allergies go west. *Nature Outlook,* 479: S2–S4. Available from: www.nature.com/nature/journal/v479/n7374_supp/full/479S2a.html

Krombach, J. W. et al. 2004. Pharaoh Menes' death after an anaphylactic reaction – the end of a myth. *Allergy* 59: 1234–1235. Available from: onlinelibrary.wiley.com/doi/10.1111/j.1398-9995.2004.00603.x/full

Lane, R. 2013. Bill Frankland: active allergist at 101. *Lancet*, Vol. 382, Issue 9894, 762. Available from: www.thelancet.com/journals/lancet/article/PIIS0140-6736%2813%2961821-8/abstract

Theoharides,T. C. et al. 2012. Mast cells and inflammation. *Biochimica et Biophysica Acta* January 2012, 1822 (1): 21–33.Available from: www.sciencedirect.com/science/article/pii/S0925443910002929

Chapter 12

D'Adamo, E. and Caprio, S. 2011. Type 2 Diabetes in Youth: Epidemiology and Pathophysiology. *Diabetes Care* May 2011, 34 (Suppl. 2): S161–S165. Available from: doi.org/10.2337/dc11-s212

Janeway, C. A. et al. 2001. Pathogens have evolved various means of evading or subverting normal host defenses. In: *Immunobiology: The Immune System in Health and Disease.* 5th edition. New York, Garland Science. Available from: www.ncbi.nlm.nih.gov/books/NBK27176/

MultiPepT1De available from: www.multipeptide.co.uk/

Redondo, M. J. et al. 2008. Concordance for Islet Autoimmunity among Monozygotic Twins. *New England Journal of Medicine* 25 December 2008, 359: 2849–2850. Available from: www.nejm.org/doi/full/10.1056/NEJMc0805398

Russell, S. J. et al. 2014. Outpatient Glycemic Control with a Bionic Pancreas in Type 1 Diabetes. *New England Journal of Medicine* 24 July 2014, 371: 313–325. Available from: www.nejm.org/doi/full/10.1056/nejmoa1314474#t=article

Chapter 13

Allenspach, E. et al. X-Linked Severe Combined Immunodeficiency. In: *GeneReviews.* 1993–2017. Seattle (WA): University of Washington, Seattle. Available from: www.ncbi.nlm.nih.gov/books/NBK1410/

Berg, L. J. 2008. The 'Bubble Boy' Paradox: An Answer That Led to a Question. *Journal of Immunology* 1 November 2008, 181 (9): 5815–5816. Available from: www.jimmunol.org/content/181/9/5815.full

Farmer, S. et al. 1948. Temporary remissions in acute leukaemia in children produced by folic acid antagonist, 4-aminopteroyl-glutamic acid (aminopterin). *New England Journal of Medicine* 3 June 1948, Vol. 238. Available from: www.nejm.org/doi/pdf/10.1056/NEJM194806032382301

Yi, Y. et al. 2011. Current Advances in Retroviral Gene Therapy. *Current Gene Therapy* June 2011, 11 (3): 218–228. Available from: www.ncbi.nlm.nih.gov/pmc/articles/PMC3182074/

Gore, M. E. 2003. Adverse effects of gene therapy: Gene therapy can cause leukaemia: no shock, mild horror but a probe. *Gene Therapy* 10, 4–4. Available from: www.nature.com/gt/journal/v10/n1/full/3301946a.html

Hacein-Bey-Abina, S. et al. 2003. LMO2-Associated Clonal T Cell Proliferation in Two Patients after Gene Therapy for SCID-X1. *Science* 17 October 2003, Vol. 302, Issue 5644, 415–419. Available from: www.sciencemag.org/content/302/5644/415.long

Hacein-Bey-Abina, S. et al. 2014. A Modified γ-Retrovirus Vector for X-Linked Severe Combined Immunodeficiency. *New England Journal of Medicine* 9 October 2014, 371: 1407–1417. Available from: www.nejm.org/doi/pdf/10.1056/NEJMoa1404588

Chapter 14

Chiocca, E. A. and Rabkin, S. D. 2014. Oncolytic Viruses and Their Application to Cancer Immunotherapy. *Cancer Immunology Research* April 2014, 2 (4): 295–300. Available from: cancerimmunolres.aacrjournals.org/content/2/4/295.full

Cuzick, J. et al. 2014. Estimates of benefits and harms in prophylactic use of aspirin in the general population. *Annals of Oncology* 26 (1): 47–57. Available from: annonc.oxfordjournals.org/content/26/1/47.full#T3

European Cancer Congress 2015. ECC 2015 press release: Post diagnosis aspirin improves survival in all gastrointestinal cancers. European Society for Medical Oncology. Available

from: www.esmo.org/Conferences/Past-Conferences/
European-Cancer-Congress-2015/News/
Post-Diagnosis-Aspirin-Improves-Survival-in-all-
Gastrointestinal-Cancers

Great Ormond Street Hospital for Children 2015. World first use of gene-edited immune cells to treat 'incurable' leukaemia. Available from: www.gosh.nhs.uk/news/press-releases/2015-press-release-archive/world-first-use-gene-edited-immune-cells-treat-incurable-leukaemia

Hanahan, D. and Weinberg, R. A. 2000. The Hallmarks of Cancer. *Cell* 7 January 2000, Vol. 100, Issue 1, 57–70. Available from: www.cell.com/cell/fulltext/S0092-8674(00)81683-9

Hanahan, D. and Weinberg, R. A. 2011. Hallmarks of Cancer: The Next Generation. *Cell* 4 March 2011, Vol. 144, Issue 5, 646–674. Available from: www.cell.com/cell/fulltext/S0092-8674(11)00127-9

Infuso, A. et al. 2006. European survey of BCG vaccination policies and surveillance in children 2005. *Eurosurveillance* 1 March 2006, Vol. 11, Issue 3. Available from: www.eurosurveillance.org/ViewArticle.aspx?ArticleId=604

Mittal, D. et al. 2014. New insights into cancer immunoediting and its three component phases – elimination, equilibrium and escape. *Current Opinion in Immunology* April 2014, 27: 16–25. Available from: www.ncbi.nlm.nih.gov/pmc/articles/PMC4388310/

Muehlenbachs, A. et al. 2015. Malignant Transformation of Hymenolepis nana in a Human Host. *New England Journal of Medicine* 5 November 2015, 373: 1845–1852. Available from: www.nejm.org/doi/full/10.1056/NEJMoa1505892#t=abstract

Tebas, P. et al. 2014. Gene Editing of CCR5 in Autologous CD4 T Cells of Persons Infected with HIV. *New England Journal of Medicine* 6 March 2014, 370: 901–910. Available from: www.nejm.org/doi/full/10.1056/NEJMoa1300662

Chapter 15

Baize, S. et al. 2014. Emergence of Zaire Ebola Virus Disease in Guinea. *New England Journal of Medicine* 9 October 2014, 371: 1418–1425. Available from: www.nejm.org/doi/full/10.1056/NEJMoa1404505#t=article

Bausch, D. G. and Schwarz, L. 2014. Outbreak of Ebola Virus Disease in Guinea: Where Ecology Meets Economy. *PLoS Neglected Tropical Diseases*, 8(7): e3056. Available from: journals.plos.org/plosntds/article?id=10.1371/journal. pntd.0003056

Henao-Restrepo, A. M. et al. 2015. Efficacy and effectiveness of an rVSV-vectored vaccine expressing Ebola surface glycoprotein: interim results from the Guinea ring vaccination cluster-randomised trial. *Lancet* 29 August 2015, Vol. 386, Issue 9996, 857–866. Available from: thelancet.com/journals/ lancet/article/PIIS0140-6736(15)61117-5/fulltext

Keim, P. et al. 2001. Molecular Investigation of the Aum Shinrikyo Anthrax Release in Kameido, Japan. *Journal of Clinical Microbiology* December 2001, 39 (12): 4566–4567. Available from: www.ncbi.nlm.nih.gov/pmc/articles/PMC88589/

Mabey, D. et al. 2014. Airport screening for Ebola. *BMJ* 14 October 2014, 349. Available from: www.bmj.com/content/349/bmj. g6202

Mann, J. G. 2012. Anthrax toxin protective antigen – Insights into molecular switching from prepore to pore. *Protein Science* January 2012, 21 (1): 1–12. Available from: www.ncbi.nlm. nih.gov/pmc/articles/PMC3323776/

World Health Organisation. *Ebola outbreak 2014–2015.* Available from: www.who.int/csr/disease/ebola/en/

Chapter 16

Abedon, S. T. et al. 2011. Phage treatment of human infections. *Bacteriophage* March–April 2011, 1 (2): 66–85. Available from: www.ncbi.nlm.nih.gov/pmc/articles/PMC3278644/

Coates, A. R. M. et al. 2011. Novel classes of antibiotics or more of the same? *British Journal of Pharmacology* May 2011, 163 (1): 184–194. Available from: www.ncbi.nlm.nih.gov/pmc/ articles/PMC3085877/

Friedman, R. M. 2008. Clinical uses of interferons. *British Journal of Clinical Pharmacology* February 2008, 65 (2): 158– 162. Available from: www.ncbi.nlm.nih.gov/pmc/articles/ PMC2253698/

Ling, L. L. et al. 2015. A new antibiotic kills pathogens without detectable resistance. *Nature* 22 January 2015, 517: 455–459. Available from: www.nature.com/nature/journal/v517/n7535/full/nature14098.html

Liu, J. K. H. 2014. The history of monoclonal antibody development – Progress, remaining challenges and future innovations. *Annals of Medicine and Surgery* (London) December 2014, 3 (4): 113–116. Available from: www.annalsjournal.com/article/S2049-0801(14)00062-4/fulltext

West, H. 2015. Immune Checkpoint Inhibitors. *JAMA Oncology* 1 (1): 115. Available from: oncology.jamanetwork.com/article.aspx?articleid=2174768

Acknowledgements

Thanks are owed to every family member or friend who gave encouragement, sent me a bizarre immunology story, read a draft or patiently understood why I wasn't coming out to play for two years.

Special thanks are due to Jim at Bloomsbury Sigma for asking me whether I'd ever thought about writing a book – I hadn't, and I would never have started this incredible journey without him. My appreciation is also extended to every talented professional who had a hand in making my random thoughts a tangible reality, from my lovely editor and my gifted illustrators, to the beautiful bookshops that stock the book.

The biggest debt of gratitude is owed to my husband Tom, for his unending love, patience, positivity and willingness to listen to stories about everything from plague to plastic vaginas. For better, for worse, eh?

Finally thank *you* for picking this up and having a read – I am indebted to your curiosity and hope I've repaid it with a treasure trove of facts and stories.

Index

Page numbers in **bold** refer to figures.